Statistics
with the TI-83™

Gloria Barrett

meridian
CREATIVE GROUP
A DIVISION OF LARSON TEXTS, INC.

Address all correspondence to:

Meridian Creative Group
A Division of Larson Texts, Inc.
5178 Station Road
Erie, PA 16510

http://www.meridiancg.com

(800) 530-2355

Other Meridian Creative Group products:

    CBL Explorations in Calculus for the TI-82
    CBL Explorations in Calculus for the TI-85
    CBL Explorations in Precalculus for the TI-82
    CBL Explorations in Precalculus for the TI-85
    CBL Explorations in Algebra for the TI-82 and TI-83
    CBL Explorations in Biology for the TI-82 and TI-83
    CBL Explorations in Chemistry for the TI-82 and TI-83
    CBL Explorations in Physics for the TI-82 and TI-83

    Helaman Ferguson: Mathematics in Stone and Bronze

    Meridian Math Coach
        Algebra 2
        Trigonometry
        Precalculus
        Probability and Statistics

Trademark Acknowledgment: TI-83 is a trademark of Texas Instruments Incorporated.

Copyright © 1997 by Meridian Creative Group, a Division of Larson Texts, Inc.

All rights reserved. No part of this publication may be reproduced or transmitted in any form or by any means, electronic or mechanical, including photocopy, recording, or any information storage or retrieval system, without prior written knowledge of Meridian Creative Group, a Division of Larson Texts, Inc.

Printed in the United States of America.

International Standard Book Number: 1-887050-31-11

1   2   3   4   5   6   7   8   9   0

Preface

Statistics with the TI-83 is written for use in any introductory, noncalculus-based statistics course. Teachers of the AP Statistics course as well as instructors of first-year college statistics courses will find that the advanced statistical features of the TI-83 graphing calculator help students analyze data, perform simulations, and develop understanding of the basic ideas presented in these courses. *Statistics with the TI-83* provides many classroom-tested examples that describe how to integrate the TI-83 into classroom instruction.

Statistics with the TI-83 is much more than a keystroke guide. This resource book consists of twenty-three chapters and three appendices. In addition to describing how to use the TI-83's statistical functions, each chapter provides activities and exercises ready for classroom use. Some activities use the graphical features of the TI-83 to analyze data. Others use the random number generators to perform simulations for solving probability problems and investigating sampling distributions. Additional activities describe how to use the TI-83's distribution and statistical features to create confidence intervals and perform significance tests.

The activities in this book can be used independently of each other; however, more detailed instructions are provided when a feature is introduced. Readers who skip around while reading this book may notice that some instructions are described in less detail than others. Most likely this occurs because the more detailed instructions are provided in a previous chapter. If more information is needed, the *TI-83 Graphing Calculator Guidebook* is recommended as a valuable source.

Because *Statistics with the TI-83* is written for statistics instructors, it is assumed that readers of this book know the underlying statistical concepts relevant to each activity. It is also assumed that readers of this book have some experience using a TI-82 or TI-83 graphing calculator. It is recommended that readers begin by reviewing the appendices where instructions are provided concerning how to use the TI-83's lists for statistical purposes.

Contents

1 Summarizing Univariate Data   1
     Calculating Summary Statistics   1
     Using Lists to Calculate Standard Deviation   3
     Using Frequency Lists   4

2 Graphing Univariate Data   7
     Graphing Modified and Regular Box Plots   8
     Graphing Histograms   10
     Graphing Box Plots and Histograms Simultaneously   12
     Summarizing and Graphing Univariate Data   13

3 Time Plots   15
     Creating a Time Plot   16

4 Investigating the Effects of Changing Units on Summary Statistics and Graphs   19
     Representing Data Sets as Percents   20
     More on the Effects of Changing Units   21
     Simulation Activity   22

5 The Normal Distribution   23
     Graphing Normal Curves   23
     Graphing and Calculating Areas Associated with Normal Curves   25
     Verifying the 68–95–99.7 Rule   26

6 Normal Distribution Calculations   27
     Calculating Areas without Graphing   27
     Calculating Percentiles   29

7 Assessing Normality; Standardized Scores   31
     Verifying a Normal Distribution   31
     Calculating Standardized Scores   34

8 Linear Bivariate Relationships   37

Creating a Scatter Plot   38
Calculating and Graphing Least-Squares Regression
   Lines   40
Making Predictions Using a Regression Line   41
Examining Residuals   42
Creating a Residual Plot   43
Calculating Correlation Coefficients   44
Investigating the Sensitivity of Regression and
   Correlation Calculations   47

9 Transformations to Achieve Linearity   51

Performing Log-Log Transformations   51
Performing Semi-Log Transformations   53

10 Using Simulations to Estimate Probabilities   57

Simulating Baseball Probabilities   57
Simulating the Birthday Problem   58

11 Binomial Probabilities   61

Creating a Binomial Probability Distribution   61
Creating a Binomial Probability Histogram   62
Determining Summary Measures for a Binomial
   Distribution   63
Approximating a Binomial Distribution   64

12 The Sampling Distribution of Sample Proportions   67

Investigating the Variability of Sample Proportions   67
Investigating the Distribution of Sample Proportions   68

13 The Sampling Distribution of Sample Means   71

Investigating a Distribution of Sample Means from
   a Normal Population   71
Investigating a Distribution of Sample Means from
   a Non-Normal Population   74

14 Inference Procedures for Means (σ known)   79

Creating a Confidence Interval for a Population Mean   79
Developing Understanding of Confidence Intervals   81
Performing a Significance Test for a Population Mean   83
Investigating the Effect of Sample Size on
   Significance Tests   86

| 15 | **Fixed Level Testing and Errors in Significance Tests** | **87** |

    Conducting Fixed Level Tests   87
    Investigating Type I Errors   89
    Investigating Type II Errors   90
    Calculating the Probability of a Type II Error   91

| 16 | **The *t* Distribution** | **93** |

    Graphing Density Curves for *t* Distributions   93
    Comparing Areas Under the *t* and Normal Curves   95
    Calculating *t* Values Associated with Percentiles   98

| 17 | **One-Sample and Matched Pairs *t* Procedures** | **101** |

    Creating a One-Sample *t* Confidence Interval   101
    Performing a One-Sample *t* Significance Test   104
    Analyzing Matched Pairs   105

| 18 | **Two-Sample *t* Procedures** | **109** |

    Performing a Two-Sample *t* Significance Test   110
    Creating a Two-Sample *t* Confidence Interval   112

| 19 | **Inference for Population Proportions** | **115** |

    Creating a Confidence Interval for a Population Proportion   115
    Developing Understanding of Confidence Intervals for Population Proportions   118
    Performing a Significance Test for a Population Proportion   120

| 20 | **Comparing Two Proportions** | **123** |

    Investigating the Sampling Distribution of $\hat{p}_1 - \hat{p}_2$ 123
    Creating a Confidence Interval for $p_1 - p_2$ 126
    Performing a Significance Test for $p_1 - p_2$ 127

| 21 | **Chi-Square Tests** | **131** |

    Graphing Density Curves for Chi-Square Distributions   131
    Calculating Areas Associated with Chi-Square Distributions   133
    Performing a Chi-Square Test for Goodness of Fit   135
    Performing a Chi-Square Test for Independence   137

22 Inference for Slope of a Regression Line   **141**

    Calculating Standard Error About a Regression Line   142
    Performing a Significance Test for β 144
    Creating a Confidence Interval for β 146

23 One-Way Analysis of Variance   **149**

    Graphing Density Curves for F Distributions   149
    Performing a One-Way ANOVA Test   151
    Creating a Confidence Interval for μ_i 153

Appendix A
Storing Data in Lists   **A1**

    Entering Data into Standard Lists   A2
    Naming a List   A3
    Using Programs to Store Lists   A5
    Deleting Lists from Memory   A6
    Using Programs to Restore Lists   A6

Appendix B
Using SetUpEditor   **B1**

    Restoring the Stat List Editor   B1
    Defining the Stat List Editor Display   B2
    Creating and Displaying New Lists   B2

Appendix C
Defining Lists Using Formulas   **C1**

    Defining Lists Using Unattached Formulas   C1
    Defining Lists Using Attached Formulas   C2
    Editing Attached Formulas   C3
    Detaching Formulas from Lists   C3

1 Summarizing Univariate Data

The activities in this chapter describe how to use the TI-83 to investigate univariate data. Specifically, in this chapter you will learn how to use the TI-83 to

- calculate summary statistics.
- use lists to calculate standard deviation.
- use frequency lists.

Calculating Summary Statistics

To calculate the summary statistics for a data set using the TI-83, use the **1-Var Stats** command. The summary statistics that can be calculated using this command include: the mean of the data, the sum of the data, the sum of the squared data, the sample standard deviation, the population standard deviation, the sample size, and the five-number summary statistics.

The following table lists the name and the length of reign (in years) for each English monarch beginning with the House of Normandy. You will use these data as you learn how to calculate summary statistics using the TI-83.

Reign of English Monarchs

NAME	REIGN	NAME	REIGN	NAME	REIGN
William I	21	Henry VI	39	William III	13
William II	13	Edward IV	22	Mary II	6
Henry I	35	Edward V	0	Anne	12
Stephen	19	Richard III	2	George I	13
Henry II	35	Henry VII	24	George II	33
Richard I	10	Henry VIII	38	George III	59
John	17	Edward VI	6	George IV	10
Henry III	56	Mary I	5	William IV	7
Edward I	35	Elizabeth I	44	Victoria	63
Edward II	20	James I	22	Edward VII	9
Edward III	50	Charles I	24	George V	25
Richard II	22	Charles II	25	Edward VIII	1
Henry IV	13	James II	3	George VI	15
Henry V	9				

Table 1.1 (Source: The World Almanac, 1996)

To calculate the summary statistics for this data set, do the following.

1. Use **SetUpEditor** to create a list named REIGN.
2. Enter the data from Table 1.1 into list REIGN.
3. Calculate the summary statistics on list REIGN by entering the following keystrokes.

    [STAT] [▶] [ENTER]   *Select* **1-Var Stats** *command.*
    [2nd] [LIST] Choose REIGN.   *Paste desired list.*
    [ENTER]

    Note: If you press [ENTER] before pasting list name REIGN after the **1-Var Stats** command, the TI-83 will calculate the summary statistics for the default list, L1.

If you entered the data and calculated the summary statistics correctly, the calculator display should be identical to the one shown below. Note the down arrow next to $n = 40$. This indicates that there is more output to be displayed. To view this output, press the down arrow key five times.

```
1-Var Stats
 x̄=21.875
 Σx=875
 Σx²=29663
 Sx=16.42572638
 σx=16.21910525
↓n=40
```

Classroom Exercises

1. Discuss the summary statistics for list REIGN. For example, compare the mean and median of the data. Why might they be different? What is the range of the data? What is the interquartile range?

2. Which standard deviation should be used in the context of this data set? Explain your reasoning. Notice that by rounding each standard deviation to the nearest integer, the two values are equal. Explain why you can expect s and σ to be approximately equal in this situation.

Using Lists to Calculate Standard Deviation

When you use the **1-Var Stats** command to calculate the summary statistics for the data in Table 1.1, you should determine that the population standard deviation, σ, is approximately 16.22. (In this situation, you know that σ is correct because the data set comprises the entire population of English monarchs since the House of Normandy who have completed their reign.) Another way to calculate the standard deviation is to use lists and the definition of standard deviation. For example, you can verify the value of σ obtained above by doing the following.

1. Use **SetUpEditor** to restore the default lists L1 through L6 in the stat list editor. Clear any data currently stored in lists L1, L2, and L3, then copy the data in list REIGN to L1. (You can use list REIGN for the computations that follow. However, using lists that can be accessed directly from the keyboard makes the computations more efficient.)

2. Define L2 to be the deviations from the mean. That is, define L2 = L1 − mean(L1). You can do this from the stat list editor by highlighting list name L2 and entering

    [2nd] [L1] [−] [2nd] [LIST] [◄] [3] [2nd] [L1] [)] [ENTER].

    You can use the value of the mean of L1 obtained when you calculated the summary statistics for list REIGN; however, to avoid rounding errors you should instruct the TI-83 to calculate the mean again.

    > Verify that the sum of the deviations from the mean is 0 by entering [2nd] [LIST] [◄] [5] [2nd] [L2] [)] [ENTER] from the home screen.

3. Define L3 to be the squared deviations. That is, define L3 = L2². You can do this from the stat list editor by highlighting list name L3 and entering

    [2nd] [L2] [x²] [ENTER].

4. To complete the computation, divide the sum of the squared deviations in L3 by the number of data entries ($n = 40$), then take the square root of the result. You can do this by evaluating the expression $\sqrt{(\text{sum}(L3)/40)}$ from the home screen. To evaluate the expression, enter

    [2nd] [√] [2nd] [LIST] [◄] [5] [2nd] [L3] [)] [÷] 40 [)] [ENTER].

    The result is 16.21910525, which is equal to the population standard deviation obtained using the **1-Var Stats** command.

Using Frequency Lists

In this section, you will learn how to use the TI-83 to organize data using frequency lists and how to calculate summary statistics on the data.

Using frequency lists enables you to efficiently organize data. Each element in a frequency list represents the number of occurrences of a corresponding entry in the data list you are analyzing. For example, if the lists L1 and L1FR (a frequency list) represent a data set and L1 = {1, 2, 3} and L1FR = {2, 5, 1}, then the data set consists of two 1's, five 2's, and one 3.

Organizing Data Using Frequency Lists

Anthropologists often search for artifacts at the remains of ancient settlements to determine when the site was inhabited and who occupied the site. The data set below lists the width (in millimeters) of 52 arrowheads that were collected at the Big Goose Creek settlement. (*Source*: Plains Anthropologist, 1983)

    10, 11, 11, 11, 11, 11, 12, 12, 12, 12, 12, 12, 12, 12, 13, 13, 13, 13, 13, 13, 13, 13, 13, 13, 13, 13, 13, 13, 13, 13, 14, 14, 14, 14, 14, 14, 14, 14, 14, 14, 14, 14, 14, 15, 15, 15, 15, 15, 15, 16, 17, 18

To enter the Big Goose Creek arrowhead data into the TI-83 using a frequency list, first use **SetUpEditor** to create two lists named BGC and BGCFR. Then access the stat list editor to enter the data into BGC and BGCFR. List BGC represents the arrowhead widths and list BGCFR represents the number of times each width occurs.

    BGC = {10, 11, 12, 13, 14, 15, 16, 17, 18}
    BGCFR = {1, 5, 8, 16, 13, 6, 1, 1, 1}

After entering the data, you should obtain a calculator display similar to the one shown below.

Calculating Summary Statistics Using Frequency Lists

The method for calculating summary statistics using frequency lists is very similar to the one used for calculating summary statistics of a standard list. To find the summary statistics for the Big Goose Creek arrowhead data, enter the following keystrokes.

[STAT] [▶] [ENTER] *Select* **1-Var Stats** *command.*
[2nd] [LIST] Choose BGC. *Paste desired list.*
[,] [2nd] [LIST] Choose BGCFR. *Paste frequency list.*
[ENTER]

If you correctly entered the data and calculated the summary statistics, the calculator display should be identical to the one shown below.

```
1-Var Stats
 x̄=13.30769231
 Σx=692
 Σx²=9328
 Sx=1.528018858
 σx=1.513255044
↓n=52
```

Classroom Exercises

1. Discuss the summary statistics for the Big Goose Creek arrowhead data. Cursor through the results to view the five-number summary. How does the median compare to the mean? What is the range of the data? What is the interquartile range?

2. From the home screen, press [2nd] [ENTER] to display the expression

   1-Var Stats ʟBGC,ʟBGCFR.

   Use the cursor keys and the [DEL] key to delete the comma and list ʟBGCFR from the expression. Press [ENTER] again and observe the resulting summary statistics. Specifically, notice the value of n. What information, if any, do these summary statistics provide?

3. The data set below lists the width (in millimeters) of 45 arrowheads collected from the Wortham Shelter settlement. (*Source*: Plains Anthropologist, 1983). Enter the data into the TI-83 using a frequency list. Then calculate the data's summary statistics. How do the arrowhead widths from Wortham Shelter compare to those from Big Goose Creek?

11, 12, 12, 12, 13, 13, 13, 13, 13, 13, 13, 13, 14, 14, 14, 14, 14, 14, 14, 14, 14, 14, 14, 14, 14, 14, 14, 14, 15, 15, 15, 15, 15, 15, 15, 15, 16, 16, 16, 16, 17, 17, 17, 18, 18

4. Use the summary statistics and the data lists to determine what percent of the measurements in each data set falls within one standard deviation of the mean. What percent in each group falls within two and three standard deviations of the mean?

2 Graphing Univariate Data

Graphs often provide information that numerical summary measures alone cannot provide. This chapter describes how to use the TI-83's graphical features to investigate univariate data. Specifically, in this chapter you will learn how to use the TI-83 to

- graph modified and regular box plots.
- graph histograms.
- graph box plots and histograms simultaneously.

To define a box plot or histogram, you must use one of three stat plot editors: Plot1, Plot2, or Plot3. Using a stat plot editor, you can turn a plot on or off, define the type of plot to be graphed, define the list names of the data to be graphed, and, depending on the type of graph, define the data mark. You can access a stat plot editor by pressing [2nd] [STAT PLOT] and choosing an editor. After choosing an editor, the calculator display will look similar to the one shown below.

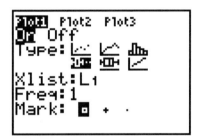

The following sections use the data listed in Table 1.1 on page 1. It is assumed that these data are stored in your calculator as list REIGN.

Graphing Modified and Regular Box Plots

To graph a modified box plot of the data in list REIGN, do the following.

1. Turn off any existing plots by entering

    [2nd] [STAT PLOT] [4] [ENTER].

2. Access Plot1 (stat plot editor) and define the modified box plot.

    Begin defining the box plot by selecting On and the modified box plot icon. (You can select each by moving the cursor over the appropriate item and pressing [ENTER].) Then paste list name REIGN after Xlist and enter 1 after Freq to indicate that each data entry in list REIGN occurs only once. Finally, select a data mark. The data mark is used for displaying outliers. After defining the box plot, the calculator display should look similar to the one shown below.

3. Define an appropriate viewing window in which to display the modified box plot.

    You can define a viewing window by pressing [WINDOW] and entering appropriate values. It is much easier, however, to use the TI-83's **ZoomStat** command. **ZoomStat** automatically defines Xmin and Xmax so the entire box plot, including outliers, is displayed. **ZoomStat** also automatically graphs the box plot. To define a viewing window and graph the box plot, enter [ZOOM] [9]. The modified box plot should look similar to the one shown below.

    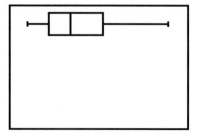

> To prevent the *x*- and *y*-axes from interfering with the box plot, turn them off by pressing [2nd] [FORMAT] and selecting the AxesOff option.

After graphing the modified box plot, use the [TRACE] features to trace the plot and determine the five-number statistics summary of the data in list REIGN.

Classroom Exercises

1. What information does the modified box plot provide? What is the minimum data value? The maximum value? What are the first and third quartiles? What is the median? Are there any outliers? Is the box plot symmetric? Discuss several reasons why the box plot is not symmetric.

2. Use **SetUpEditor** to restore the default lists L1 through L6 to the stat list editor. Clear the contents of L1 and copy the contents of REIGN into L1. Delete data entries 0 and 63 from L1. Then, define Plot2 to be a modified box plot using the data in L1 and graph Plot1 and Plot2 simultaneously. Trace and compare the two box plots.

   To move the trace cursor between box plots, press ▲ or ▼.

3. Queen Victoria reigned for 63 years. How long would she have needed to reign for the length of her monarchy to qualify as an outlier? Investigate by clearing the contents of L2, copying the contents of REIGN into L2, and sorting the data in descending order so that the longest reign (Queen Victoria's) is located at the top of the list. To sort the data in descending order, enter

   [STAT] [3] [2nd] [L2] [ )] [ENTER].

   To determine how long Queen Victoria would have needed to reign for the length of her reign to qualify as an outlier, change 63 to a larger value and use Plot3 to define and graph a modified box plot of the revised data. Use **ZoomStat,** if necessary. Repeat until you determine the shortest length of reign that qualifies as an outlier.

   Verify mathematically that the value of the outlier exceeds the upper quartile by at least 1.5 times the interquartile range.

4. Turn off Plot1. Use Plot2 to define and graph a *regular* box plot of the data in L2. Compare the result with the modified box plot obtained in Exercise 3. (Plot2 will be graphed in the upper third of the viewing window with Plot3 directly beneath it.) Trace both box plots to observe that each provides slightly different information. Discuss the information each box plot provides.

Graphing Histograms

To graph a histogram of the data in list REIGN, do the following.

1. Turn off any existing plots by entering

    [2nd] [STAT PLOT] [4] [ENTER].

2. Access Plot1 (stat plot editor) and define the histogram.

    Begin defining the histogram by selecting On and the histogram icon. (You can select each by moving the cursor over the appropriate item and pressing [ENTER].) Then paste list name REIGN after Xlist. Finally, enter 1 after Freq to indicate that each data entry in list REIGN occurs only once. After defining the histogram, the calculator display should look identical to the one shown below.

3. Define an appropriate viewing window in which to display the histogram.

    You can define a viewing window by pressing [WINDOW] and entering appropriate values. It is much easier, however, to use the TI-83's **ZoomStat** command. **ZoomStat** automatically defines Xmin, Xmax, Xscl, Ymin, and Ymax. **ZoomStat** also automatically graphs the histogram. To define a viewing window and graph the histogram, enter [ZOOM] [9]. The histogram should look similar to the one shown below.

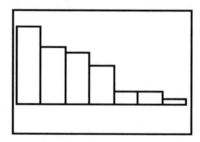

The histogram indicates that the distribution of reign lengths is skewed to the right. To obtain more detailed information, press [TRACE]. When you press [TRACE], the trace cursor appears at the top center of the left-most bar and some information is displayed on the screen. This information includes the name of the stat plot editor, the list name, the endpoints of the class interval, and the number of data values represented by the bar. The histogram bars represent all data values greater than or equal to the left endpoint and less than the right endpoint.

Classroom Exercises

1. What information does the histogram provide? How does this information correspond to the information provided by a box plot of the same data?

2. Press [WINDOW] and use the values listed below to define the window settings to create a histogram having bar widths representing 5 years. The width of each bar is determined by Xscl.

   $Xmin = -.5$ $Ymin = -2.5$

   $Xmax = 65.5$ $Ymax = 10.5$

   $Xscl = 5$ $Yscl = 1$

   After defining the window settings, press [GRAPH] to display the histogram. How many monarchs reigned less than 5 years? Less than 10 years? At least 30 years?

3. Modify the window settings to create a histogram so that each bar width represents 8 years. Then graph the histogram to verify that the histogram satisfies this requirement.

4. Compare the three histograms graphed in this section. What characteristics do they have in common? How do they differ? Which one do you think best represents the data in list REIGN? Explain your reasoning.

5. Use the Big Goose Creek arrowhead data listed on page 4 to graph a histogram. When defining the histogram, specify Xlist = BGC and Freq = BGCFR. (It is assumed that these lists are properly stored in your calculator. If not, please read the section **Using Frequency Lists.**) When defining the window settings, specify Xmin = 9.5, Xmax = 18.5, and Xscl = 1. Adjust Ymin and Ymax as necessary. Be sure to turn off other existing plots. Discuss the shape of the histogram. What conclusions can you make from the histogram?

Graphing Box Plots and Histograms Simultaneously

Graphing box plots and histograms simultaneously enables you to see how each graph indicates the skewness of the data distribution. To graph a modified box plot and a histogram of the data in list REIGN on the same viewing screen, do the following.

1. Define Plot1 to be the histogram you think best represents the REIGN data. Change Ymin and Ymax so that there is room for 2 text lines below the histogram and room for the box plot above the histogram.

2. Define Plot2 to be a box plot of the REIGN data. Because the data set has no outliers, you can use a modified or regular box plot.

3. Press [GRAPH] to display the box plot and the histogram. (If the graphs coincide, increase Ymax.) The display should be similar to the one shown below.

To trace the graphs, press [TRACE] and use the left and right arrow keys ([▶] [◀]). To switch between graphs, use the up and down arrow keys ([▲] [▼]).

Classroom Exercises

1. Find the median of the data in list REIGN by tracing the box plot. Note where the median occurs with respect to the histogram. Because the median occurs at the 50th percentile, half the area of the histogram should be on each side of the median. Does this appear to be true? Investigate by using the TI-83's drawing features to draw a vertical line on the histogram that marks the median's location. To draw the line, position the trace cursor on the median line in the box plot and press [2nd] [DRAW] [4] [ENTER].

2. Where do you expect the mean to be located relative to the median? Explain your reasoning. Draw a vertical line where you think the mean is located. Compare your guess with the actual mean by entering the following keystrokes from the home screen.

[2nd] [DRAW] [4] [2nd] [LIST] [◄] [3] ʟREIGN [)] [ENTER]

Summarizing and Graphing Univariate Data

Exercises 1–6 combine topics discussed in Chapters 1 and 2.

1. In list L1, create a list of nine test grades (based on a 100-point test) in which the mean is below the lower quartile. Use the **1-Var Stats** command to verify that your data satisfy this requirement. Then represent the data in a modified box plot.

2. In list L2, create a list of eight test grades in which the mean and median are equal. Represent the data using a modified box plot. What is the primary difference between this boxplot and the one created in Exercise 1?

3. In list L3, create a data set consisting of fifteen to twenty test grades with mean 70. Find the median, range, interquartile range, and standard deviation for these grades. Represent these data in a histogram. Is it symmetric or skewed?

4. Use **SortA** to order the grades created in L3 from smallest to largest and copy the results into list L4. Delete the four most central grades. Compute and analyze the summary statistics for the revised data. How do the results compare to those obtained in Exercise 3?

5. Copy L3 into list L5 and delete the two lowest and the two highest grades. Compute and analyze the summary statistics for the revised data. How do the results compare to those obtained in Exercise 3?

6. Use Plot1, Plot2, and Plot3 to create box plots of the data in L3, L4, and L5, respectively. Compare the box plots. How are they similar? How are they different?

3 Time Plots

In statistical analysis, you often need to determine how a variable changes with respect to time. For example, data consisting of population counts, production measurements, or growth and decay are frequently gathered and analyzed over a period of time. One way to analyze these data is to represent them using a two-dimensional graph called a time plot. In this chapter, you will learn how to create such a plot using the TI-83.

The following table lists the year and the percent of U. S. residents who say they were born in a foreign country. You will use these data as you learn how to create a time plot.

Year	Percent Foreign Born
1900	13.6
1910	14.7
1920	13.2
1930	11.6
1940	8.8
1950	6.9
1960	5.4
1970	4.8
1980	6.2
1990	7.9

Table 3.1 (Source: U.S. Bureau of the Census)

Creating a Time Plot

To create a time plot using the data in Table 3.1, do the following.

1. Use **SetUpEditor** to create lists YEAR and FORBN and to remove all other lists from the stat list editor. Also, turn off any existing stat plots.

2. Enter the data in Table 3.1 into lists YEAR and FORBN.

    An efficient way to enter the data into list YEAR is to use the TI-83's sequence command, **seq**. Consider the year data as a sequence that begins with 1900, increases by 10, and ends with 1990. To enter the data using the **seq** command, first highlight list name YEAR in the stat list editor, then enter the following keystrokes.

Keystrokes	Description
[2nd] [LIST] [▶] [5]	*Access* **seq** *command.*
[X,T,Θ,n] [,]	*Define expression.*
[X,T,Θ,n] [,]	*Define variable.*
1900 [,]	*Define start value.*
1990 [,]	*Define end value.*
10 [)]	*Define variable increment.*
[ENTER]	

3. Access Plot1 and define the time plot.

    Begin defining the time plot by selecting On and the time plot icon. (You can select each by moving the cursor over the appropriate item and pressing [ENTER].) Then paste list names YEAR and FORBN after Xlist and Ylist, respectively. Finally, select a data mark. After defining the time plot, the calculator display should look similar to the one shown below.

Hint: To access a list name near the bottom of the list names menu, press the [▲] key. This enables you to scroll up from the bottom of the menu. To access a list name in a more central menu location, press [ALPHA] [▼]. This allows you to scroll the menu screen by screen.

4. Define an appropriate viewing window in which to display the time plot.

   You can define a viewing window by pressing WINDOW and entering appropriate values. It is much easier, however, to use the TI-83's **ZoomStat** command. **ZoomStat** automatically defines Xmin, Xmax, Ymin, and Ymax so that the entire time plot is displayed. **ZoomStat** also automatically graphs the time plot. To define a viewing window and graph the time plot, enter ZOOM 9. The time plot should look similar to the one shown below.

   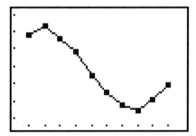

Classroom Exercises

1. What information does the time plot provide? Use the TI-83's trace feature to examine the time plot.

2. Information obtained from a time plot often depends on the scale used for each axis. Press WINDOW to see how the window settings are defined using **ZoomStat.** Because the vertical axis represents percent values which can range from 0 to 100, it seems reasonable to define Ymin=0 and Ymax=100. Make these changes and examine the new graph. How does it compare to the original one? Are the trends still apparent? Do the changes over time appear smaller or larger on the new graph?

3. Using the TI-83, you can graph as many as three time plots simultaneously. This helps when comparing and identifying trends in related data sets. Use a resource almanac to research the men's and women's winning times for an Olympic event. In lists L1, L2, and L3, enter the Olympic years, the men's winning times, and the women's winning times, respectively. (Use **seq** to enter the Olympics years.) Create a time plot to represent each data set. How have the winning times changed with respect to time? Which group has improved the most? Are both groups still improving? How has the difference between men's and women's winning times changed?

4 Investigating the Effects of Changing Units on Summary Statistics and Graphs

The activities in this chapter describe how to use the TI-83's list features to investigate the effect of changing units on graphs and summary statistics. Specifically, you will investigate the effect of changing units on summary statistics and graphs when

- a data set is represented as percents.
- a data set increases by a constant.
- a data set increases by a percent.
- a data set increases by a percent and also increases by a constant.

Also, a simulation activity is included at the end of the chapter.

The following test scores are the results from one class on a 150-point test. You will use these data as you investigate the effect of changing units on summary statistics and graphs.

122, 133, 138, 123, 125, 119, 108, 114, 117, 125, 92, 126, 125, 110, 133, 132, 124, 115, 120, 126, 122, 127, 113, 142, 129, 130

Use **SetUpEditor** to restore the stat list editor to the standards lists, L1 through L6. Clear any data currently in L1, L2, and L3. Then enter the test scores into L1.

Classroom Exercises

1. Use **1-Var Stats** to determine the mean, median, and standard deviation of the data in L1. Also, determine the range and the interquartile range. Record these summary statistics.

2. Use Plot1 to define and graph a histogram of the test scores in L1. Describe the histogram's shape. Does the data distribution appear symmetric? Are there any gaps or outliers? If so, identify them.

3. Use Plot2 to define and graph a modified box plot of the test scores in L1. Discuss the box plot's shape. Are there any outliers? If so, identify them.

4. Copy the test scores in L1 into L2, then delete any outliers from L2. Use **1-Var Stats** to determine the summary statistics for list L2. What effect does deleting the outlier have on the summary statistics? That is, which measure of central tendency, the mean or the median, is most affected by the outlier? Which measure of spread, the standard deviation or the interquartile range, is most affected by the outlier? Explain your reasoning.

5. Define and graph a histogram and a modified box plot of the data in L2. How do the new graphs compare to those obtained in Exercises 2 and 3? Discuss the effect that deleting the outlier has on the graphs.

Representing Data Sets as Percents

Hint: The TI-83's **SortD** command allows you to search a list efficiently. To sort list L1 in descending order, enter
[STAT] [3] [2nd] [L1] [)] [ENTER].

Suppose grades are based on a ten percent point scale. For example, a grade of 90% or higher is considered an A, a grade of 80% to 89% is considered a B, and so forth. A student, therefore, would need a score of $150 \times 0.9 = 135$ points to earn an A on the test. Search list L1 to determine how many students earned an A.

Because grades are based on percents, it is more convenient to analyze the test results by converting the test scores in L1 to percents. To convert the test scores to percents and store the results in L2, define $L2 = (L1/150) \times 100$. (Note that you are simply multiplying the original test scores by 2/3). Examine the contents of L2 to verify that only two students earned an A. How many students earned a B? C? D?

Classroom Exercises

1. How can you determine the mean and the median of the percent scores using the original test scores? Investigate by calculating the summary statistics for L2 and comparing the results to those obtained for L1. Record the summary statistics for L2.

2. Compare the summary statistics for L2 to those obtained for L1. Specifically, discuss the relationship between the standard deviation of the percent scores in L2 and the original scores in L1. How did the conversion affect the minimum and maximum grades? How did it affect the lower and upper quartiles? The range and the interquartile range?

3. Define and graph a box plot and a histogram of the percent scores in L2. Compare each graph to the original box plot and histogram. Do the shapes of the graphs change? If so, explain how.

4. Explain how to define the window settings for the histogram so that each grade, A, B, C, and D, is represented separately. Verify your results by tracing the histogram and determining the number of grades represented by each histogram bar.

More on the Effects of Changing Units

Suppose the instructor is disappointed that only two students earned an A on the test and is considering ways to curve the grades. To curve the test grades, the instructor considers three options.

1. Increasing the percent scores by adding 2% points to each score
2. Increasing the percent scores by 3%
3. Increasing the percent scores by 2%, then adding 1% point to each score

In the following exercises, you will investigate the effect of these changes on the summary statistics and graphs.

Classroom Exercises

1. To increase the percent scores by adding 2% points to each score and store the results in list L3, define L3 = L2 + 2. What effect does adding 2% points to each score have on the summary statistics? Do any summary measures remain the same? Which measures change?

    What effect does adding 2% points to each score have on the graphs? Investigate by creating a histogram and a modified box plot of the data in L3. How does the histogram change? Does your response depend on the class intervals? Explain. How does the modified box plot change?

2. To increase the percent scores by 3% and store the results in L3, define L3 = 1.03 × L2. Investigate the effect this change has on the summary statistics and graphs.

3. To increase the percent scores by 2% and then add 1% point to each score, define L3 = 1.02 × L2 + 1. Investigate the effect this change has on the summary statistics and graphs.

Simulation Activity

In this activity, you will use the TI-83's random integer generator **randInt** to simulate tossing a fair, six-sided die one hundred times.

Before beginning the activity, instruct each student to define a random seed value. Doing so allows each student to generate a different data set. A possible seed value can be the last four digits of a student's telephone number. To define a seed value, enter the following keystrokes.

Seed value [STO▶] [MATH] [◀] [1] [ENTER]

To simulate the roll of a die using **randInt**, enter the following keystrokes.

[MATH] [◀] [5]   Access **randInt** command.
1 [,]   Define lower bound.
6 [)]   Define upper bound.
[ENTER]

After pressing [ENTER], notice that the calculator displays a number between 1 and 6, inclusive. To "roll" the die again, simply press [ENTER]. You can obtain 100 rolls by repeatedly pressing [ENTER] and storing each result in a list. A more efficient way to simulate and record the 100 rolls, however, is to execute the following command from the home screen.

randInt(1,6,100) → L1

This command rolls the die one hundred times and stores the results in list L1.

Classroom Exercises

1. Calculate the summary statistics for the data in L1. Record the results.

2. Use Plot1 to define a histogram of the data in L1. Graph the histogram using the following window settings: Xmin = .5, Xmax = 6.5, Xscl = 1, Ymin = 0, Ymax = 30, Yscl = 5. Does the distribution appear reasonably uniform? Explain.

3. Investigate the effect each change listed below has on the summary statistics and histogram. Anticipate the results before actually calculating the summary statistics and graphing the histogram. Store the altered data in L2. Change window settings as necessary.

   (a) Multiply L1 by 10.   (d) Add 100 to L1.
   (b) Multiply L1 by −1.   (e) Subtract 7 from L1.
   (c) Divide L1 by 10.   (f) Multiply L1 by 2, then add 10.

5 The Normal Distribution

Normal probability density functions are usually the first continuous density functions studied in introductory statistics courses. In this chapter, you will learn how to use the TI-83 to investigate graphs and properties associated with normal distributions. Specifically, you will learn how to use the TI-83 to

- graph normal curves.
- graph and calculate areas associated with normal curves.
- verify the 68–95–99.7 Rule.

Graphing Normal Curves

Consider a normal probability density function with mean 80 and standard deviation 5. To graph a normal curve for this distribution, do the following.

1. Turn off or clear any existing Y= variables, stat plots, and drawings.

2. Access the Y= editor and define the normal probability density function by entering the following keystrokes.

    [Y=]   *Access* Y= *editor.*
    [2nd] [DISTR] [1]   *Access* **normalpdf** *command.*
    [X,T,Θ,n] [,] 80 [,] 5 [)]   *Enter function.*

3. Define an appropriate viewing window.

    Because the function is a normal probability density function with mean 80, you can expect the graph to be symmetric and to have a maximum value at $x = 80$. Therefore, define Xmin = 60, Xmax = 100, and Xscl = 10. Also, because density curves cannot have negative values, define Ymin = 0. Finally, define Ymax by evaluating the density function at the mean and storing the result in Ymax. To define Ymax, cursor to Ymax and enter

    [2nd] [DISTR] [1] 80 [,] 80 [,] 5 [)] [ENTER].

    The resulting value is approximately 0.07978845.

The Normal Distribution   23

4. Press GRAPH.

After pressing GRAPH, you should obtain a display similar to the one shown below. Trace the graph to verify that the maximum value occurs when $x = 80$. Note that when you press TRACE, the screen text interferes with the graph. To make a friendlier viewing window, define Ymin $= -.01$ and Ymax $= .09$.

Classroom Exercises

1. Trace each side of the graph to find a y-value less than 0.001. What are the corresponding x-values? How far from the mean do these x-values occur?

2. The points at which a graph's curvature changes are called inflection points. The graph of a normal probability density function has two inflection points. Trace the graph to estimate the x-value of each inflection point. How far are these values from the mean?

3. Press 2nd [CALC] 1. The graph, along with a prompt for an x value, is displayed. Enter 85. The trace cursor should be positioned at the point on the graph where $x = 85$. What is the significance of this point? What is the relationship between this point and the inflection points determined in Exercise 2?

4. Use the Y= editor to define and graph a normal probability density function with mean 80 and standard deviation 7. Enter the function in Y2. Also, change the graph style by cursoring to the left of Y2 and pressing ENTER. Notice that the graph line changes to a thick line. Press GRAPH to graph both the original and the new normal probability density functions. Compare the graphs. How are they similar? How are they different?

5. Clear the function in Y2. Then enter and graph a normal probability density function with mean 84 and standard deviation 5. Compare the resulting graph to the original graph. How are they similar? How are they different?

Graphing and Calculating Areas Associated with Normal Curves

The total area under a normal curve is always equal to one. To verify that the total area under the graph of the normal probability density function with mean 80 and standard deviation 5 is equal to one, do the following.

1. Clear any normal probability density functions from the Y= editor.

2. Define an appropriate viewing window. (See page 23, if necessary.)

3. To draw the graph and calculate the area of the region under the graph between $x = -1000$ and $x = 1000$, enter the following keystrokes.

   [2nd] [DISTR] [▶] [1]   *Access* **ShadeNorm** *command.*
   -1000 [,] 1000 [,]   *Define lower and upper bounds.*
   80 [,] 5 [)]   *Define mean and standard deviation.*
   [ENTER]

   After pressing [ENTER], you should obtain a display similar to the one shown below. Notice that the area under the curve is equal to one.

To clear a drawing from the home screen, enter [2nd] [DRAW] [1] [ENTER]. To clear a drawing from the graphing screen, enter [2nd] [DRAW] [1].

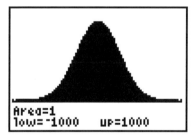

Classroom Exercise

1. Verify that the area of the region between the mean and $x = 1000$ is equal to 0.5.

Verifying the 68–95–99.7 Rule

In a normal distribution with mean μ and standard deviation σ, 68% of the area under the normal curve falls within σ of the mean, 95% falls within 2σ, and 99.7% falls within 3σ. To verify the 68–95–99.7 Rule for the normal distribution $N(80, 5)$ using the TI-83, execute the following commands from the home screen. Be sure to clear any drawings before executing each command.

ShadeNorm(75, 85, 80, 5):

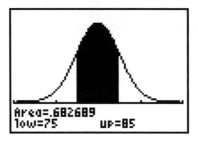

ShadeNorm(70, 90, 80, 5):

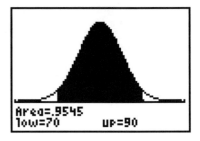

ShadeNorm(65, 95, 80, 5):

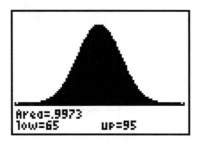

Classroom Exercise

1. Verify that the 68–95–99.7 Rule holds for the following normal probability density functions.

   a. $N(80, 7)$ b. $N(84, 5)$

   c. $N(0, 1)$

6 Normal Distribution Calculations

In Chapter 5, you learned how to graph normal curves and calculate the area of a region under a normal curve. In this chapter, you will learn how to use the TI-83 to

- calculate areas without graphing.
- calculate percentiles.
- solve for an unknown mean.

The problems posed in this chapter are typical of those found in most introductory statistics texts. However, each solution describes how to solve the problem using the TI-83's statistical features rather than using the more traditional standard normal tables.

Calculating Areas without Graphing

Mr. Goebel, a statistics instructor, can drive two different routes to school. He usually leaves home at 7:20 so that he can be at school for an 8:00 class. Last year he compared the travel times for each route by recording how long it took him to get from home to school each day. One week he would travel route A; the next week he would travel route B. At the end of the year he summarized and graphed the results for each route. Both sets of travel times appeared normally distributed. The mean travel time for route A was 35 minutes with a standard deviation of 3 minutes while the mean time for route B was 30 minutes with a standard deviation of 5 minutes.

Classroom Exercises

1. Graph the normal probability density functions, $N(35, 3)$ and $N(30, 5)$, in the same viewing window. Compare the graphs. How are they similar? How are they different?

2. Use **ShadeNorm** to determine the proportion of the trips along route A that take longer than 30 minutes. What proportion of the trips along route B take longer than 30 minutes?

Calculating Areas

Mr. Goebel wants to determine which travel route, A or B, provides the greater percent of on-time arrivals. Because he leaves home at 7:20 and needs to be at school by 8:00, the better route is the one that provides the greater percent of travel times that take less than 40 minutes. Which route is better?

To solve this problem, you can use the **ShadeNorm** command as shown in Chapter 5. A more efficient method for solving this problem, however, is to use the **normalcdf** command. Given a mean μ and a standard deviation σ, **normalcdf** calculates the normal distribution area between a specified lower and upper bound.

To determine the percent of travel times for route A that take less than 40 minutes, do the following.

2nd [DISTR] 2	*Access **normalcdf** command.*
0 , 40 ,	*Define lower and upper bounds.*
35 , 3)	*Define mean and standard deviation.*
ENTER	

The result is approximately 0.9522. This means that about 95% of the time, it will take Mr. Goebel less than 40 minutes to get to school via route A.

Classroom Exercise

1. With respect to travel time, is route B a better route than route A? Investigate by finding the percent of travel times for route B that take less than 40 minutes.

Calculating Percentiles

Mr. Goebel really prefers to travel along route A, so he is disappointed that route B seems slightly better. He wonders how much earlier he would have to leave to have a 98% on-time rate if he travels route A. (He assumes that travel times for slightly earlier departures are normally distributed with mean 35 minutes and standard deviation 3 minutes.) At what time should Mr. Goebel depart to have a 98% on-time rate if he travels route A?

To solve this problem, you can use the TI-83's **invNorm** command. Given a mean μ and a standard deviation σ, **invNorm** calculates the inverse cumulative normal distribution function for a given area under the normal curve. You can determine the 98th percentile for $N(35, 3)$ by entering the following keystrokes.

[2nd] [DISTR] [3]	*Access* **invNorm** *command.*
0.98 [,]	*Define percentile.*
35 [,] 3 [)]	*Define mean and standard deviation.*
[ENTER]	

The result is approximately 41.16. This means that to obtain a 98% on-time rate, Mr. Goebel must allow for a travel time of slightly more than 41 minutes when he travels route A. He would need to depart slightly before 7:19.

Solving for an Unknown Mean

Part of Mr. Goebel's trip along route B consists of travel on an interstate highway. Next year the speed limit on the highway will be increased, so Mr. Goebel is hoping that he will be able to reduce the mean time that it takes him to get to school along route B. (He expects the standard deviation to remain at 5 minutes.) What must the mean travel time be so that Mr. Goebel could depart from home at 7:30, travel along route B, and arrive at school on time 98% of the time?

This problem differs from previous ones because the normal probability density function is unknown. You know that the maximum time value is 30 minutes, the probability is 0.98, and the standard deviation is 5; however, you do not know the mean. To solve this problem, use the TI-83's equation solver.

To use the equation solver to find the mean, enter the following keystrokes.

Keystrokes	Description
MATH 0	*Display equation solver screen.*
▲ CLEAR	*Clear equation solver screen, if necessary.*
2nd [DISTR] 2	*Paste **normalcdf** command.*
0 , 30 , X ,	*Enter equation: lower and upper bounds, mean*
5) − .98	*variable, standard deviation, percentile*
ENTER	
25	*Enter initial guess.*
ALPHA [SOLVE]	*Solve equation.*

> When prompted to make an initial guess, make as reasonable a guess as possible. Be sure the cursor is on the X= entry line before pressing ALPHA [SOLVE].

The resulting x-value is approximately 19.727. This means that Mr. Goebel's mean travel time must be slightly less than 20 minutes if he wants to depart from home at 7:30, travel route B, and be on time 98% of the time.

Classroom Exercises

1. Discuss the feasibility of the solution obtained above. What must the mean travel time be if Mr. Goebel wants to leave home at 7:30 and have an on-time rate of 95%?

2. Suppose Mr. Goebel wants to depart at 7:25 and travel along route B. What must the mean travel time be to obtain an on-time rate of 95%? 98%?

Assessing Normality; Standardized Scores

The activities in this chapter describe how to use the TI-83 to

- verify whether a data set is normally distributed.
- calculate standardized scores.

The following data are the serum cholesterol levels of fifty men, aged 38 to 41 years, who participated in a study done several years ago at the University of North Carolina in Chapel Hill. Use **SetUpEditor** to create a list named CHOL. Then enter the data into list CHOL.

285, 267, 210, 248, 202, 339, 291, 303, 250, 198, 218, 235, 303, 197, 187, 191, 157, 266, 230, 208, 191, 234, 263, 238, 175, 226, 228, 140, 254, 221, 253, 231, 222, 246, 262, 201, 259, 211, 263, 165, 160, 217, 122, 178, 212, 245, 183, 179, 251, 220

Verifying a Normal Distribution

Variables that occur in nature and industry, as well as repeated measurements of the same quantity, are often normally distributed. One way to determine whether a data set *might* be normally distributed is to examine a histogram of the data. If the histogram appears symmetric and mound-shaped, much like a normal curve, then the data are likely normally distributed. To verify whether the data are normally distributed, you can

- use the 68–95–99.7 Rule.
- create a normal probability plot.

> Use Plot1 to define and graph a histogram. (When defining the window settings, let Xmin = 100, Xmax = 360, and Xscl = 20.) Do the cholesterol data in list CHOL appear to be normally distributed? Explain.

Using the 68–95–99.7 Rule to Verify Normality

To verify that a data set is normally distributed using the 68–95–99.7 Rule, do the following.

1. Use **1-Var Stats** to determine the mean and standard deviation of the data.

    For the data in list CHOL, you should obtain a mean of 224.7 and a standard deviation of approximately 43.56.

2. Show that approximately 68% of the sample data fall within one standard deviation of the mean.

   With respect to list CHOL, you must determine the percent of men having cholesterol levels between 224.7 − 43.56 = 181.14 and 224.7 + 43.56 = 268.26. To determine the number of men having such cholesterol levels, sort the data using **SortA** and count the appropriate data entries. You should determine that the cholesterol levels of 37 men, or 74% of the sample data, fall within one standard deviation of the mean.

3. Show that approximately 95% of the sample data fall within two standard deviations of the mean.

4. Show that approximately 99.7% of the sample data fall within three standard deviations of the mean.

5. Make a conclusion based on the results of Parts 2–4.

Classroom Exercises

1. Do approximately 95% of the data in list CHOL fall within two standard deviations of the mean? Explain.

2. Do approximately 99.7% of the data in list CHOL fall within three standard deviations of the mean? Explain.

3. Based on your results, are the data in list CHOL normally distributed? Explain your reasoning.

Creating a Normal Probability Plot

Using a normal probability plot provides a more sensitive assessment of the appropriateness of a normal model than simply using the 68–95–99.7 Rule. A normal probability plot graphs each data value with the standard normal z-value that corresponds to its quantile location. If the points in a normal probability plot lie in a relatively straight line, then the plot indicates that the data are normally distributed.

To graph a normal probability plot of the data in list CHOL, do the following.

1. Turn off any existing plots by entering

   [2nd] [STAT PLOT] [4] [ENTER].

2. Define the normal probability plot using Plot1 (stat plot editor).

Begin defining the plot by selecting On and the normal probability plot icon. (You can select each by moving the cursor over the appropriate item and pressing [ENTER].) Then paste list name CHOL after Data List and select X after Data Axis. Selecting X after Data Axis instructs the TI-83 to plot the data on the x-axis and the corresponding z-scores on the y-axis. Finally, select a data mark. After defining the normal probability plot, the calculator display should look similar to the one shown below.

3. Define an appropriate viewing window in which to display the normal probability plot.

You can define a viewing window by pressing [WINDOW] and entering appropriate values. It is much easier, however, to use the TI-83's **ZoomStat** command. **ZoomStat** automatically defines the viewing window so that the entire normal probability plot is displayed. **ZoomStat** also automatically graphs the plot. To define a viewing window and graph the normal probability plot, enter [ZOOM] [9]. The plot should look similar to the one shown below.

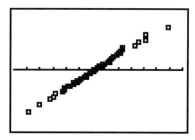

In the normal probability plot shown above, the cholesterol levels are plotted on the x-axis and the corresponding z-scores are plotted on the y-axis. Because the data points appear to lie in a straight line, you can conclude that the data are approximately normally distributed.

Classroom Exercises

1. Press [TRACE] to determine the coordinates of the left-most point in the normal probability plot. Use **invNorm (.01)** to verify that the z-value is associated with the first percentile of a $N(0, 1)$ distribution.

2. Access the stat list editor and define L1 and L2 as shown below.

    L1 = seq(X,X,.01,.99,.02)
    L2 = seq(invNorm(L1(X),X,1,50,1).

    L1 will contain the quantiles associated with the data in list CHOL and L2 will contain the z-values that correspond to the quantiles in L1. Define Plot 2 to be a scatter plot with Xlist = CHOL and Ylist = L2. Select a different mark than the one used in the normal probability plot. Graph Plot1 and Plot2 to see that they are identical.

3. Table 1.1 on page 1 lists the name and length of reign (in years) for each English monarch beginning with the House of Normandy. Use each method discussed in this chapter to determine whether these data are normally distributed.

Calculating Standardized Scores

A standardized score indicates how many standard deviations a particular data value is above or below the mean. To calculate a standardized score for a particular data value, subtract the mean from the data value and divide the result by the standard deviation. The resulting standardized score is positive if the data value is greater than the mean and negative if the data value is less than the mean.

Raw data are often converted to standardized scores to enhance understanding and to facilitate comparisons between data measured using different scales. For example, consider the cholesterol data in list CHOL. The mean cholesterol level of the fifty men is 224.7 with a standard deviation of 43.56. Therefore, a cholesterol level of 181 is one standard deviation below the mean and a level of 355 is three standard deviations above the mean. To a person untrained in the medical field but knowledgeable about statistics, it may be more informative to know that the cholesterol level is 3 standard deviations above the mean than to know the actual cholesterol level of 355.

The TI-83's list features allow you to easily convert a data set to standardized scores. To convert the cholesterol data in list CHOL to standardized scores and store the scores in a list named STDSC, do the following.

1. Use **SetUpEditor** to display list names CHOL and STDSC in the stat list editor.

2. Define list STDSC by attaching the following formula.

   STDSC = "(LCHOL − mean(LCHOL))/stdDev(LCHOL)"

   Note: Commands **mean** and **stdDev** are located in the list math menu.

Classroom Exercises

1. Cursor through list STDSC to observe the range of standardized scores. Identify the standardized scores associated with the lowest and the highest cholesterol levels.

2. Use **1-Var Stats** to calculate the mean and standard deviation of the standardized scores. Do these values surprise you? Use your knowledge of how changing units affects summary measures to explain your results.

3. Create a histogram of the standardized scores in list STDSC. How does it compare with the histogram of the original data values?

8 Linear Bivariate Relationships

In this chapter, you will learn how to use the TI-83 to investigate linear bivariate relationships. Specifically, you will learn how to use the TI-83 to

- create scatter plots.
- calculate and graph regression lines.
- make predictions using regression lines.
- examine residuals.
- create residual plots.
- calculate correlation coefficients.
- investigate the sensitivity of regression and correlation calculations.

The following table lists the name, total minutes played, and the total points scored for each player who played at least 600 minutes for the 1994 Chicago Bulls. You will use these data throughout the chapter.

1994 Chicago Bulls

PLAYER	MINUTES	POINTS
Pippen	2759	1587
Grant	2570	1057
Armstrong	2770	1212
Kukoc	1808	814
Kerr	2036	709
Myers	2030	650
Williams	638	289
Wennington	1371	542
Longley	1502	528
Cartwright	780	235
Blount	690	198

Table 8.1 (Source: 1995 World Almanac)

Use **SetUpEditor** to create and display lists MINS and PTS in the stat list editor. Then enter the data from Table 8.1 into these lists.

Creating a Scatter Plot

An effective way to examine the relationship between two variables is to create a scatter plot. To create a scatter plot of the data in lists MINS and PTS, do the following.

1. Turn off any existing plots by entering

    [2nd] [STAT PLOT] [4] [ENTER].

2. Access Plot1 (stat plot editor) and define the scatter plot.

    Begin defining the scatter plot by selecting On and the scatter plot icon. (You can select each by moving the cursor over the appropriate item and pressing [ENTER].) Then paste list names MINS and PTS after Xlist and Ylist, respectively. Finally, choose a data mark. After defining the scatter plot, the calculator display should look similar to the one shown below.

3. Define an appropriate viewing window in which to display the scatter plot.

    You can define a viewing window by pressing [WINDOW] and entering appropriate values. It is much easier, however, to use the TI-83's **ZoomStat** command. **ZoomStat** automatically defines the viewing window settings and graphs the scatter plot. To define a viewing window and graph the scatter plot, enter [ZOOM] [9]. The scatter plot should look similar to the one shown below.

    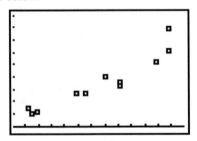

    The scatter plot indicates an increasing relationship which appears to be fairly linear.

Classroom Exercises

1. Trace the scatter plot to determine the player represented by each point. Do any points appear to be outliers? Explain.

2. The following table lists the name, total minutes played, and the total points scored for each player who played at least 600 minutes for the 1996 Chicago Bulls. Enter the data in Table 8.2 into lists named NMINS and NPTS. Then use Plot2 to define and create a scatter plot of these data. (Turn off Plot1.) Describe the relationship between total points scored and total minutes played for these players. Trace the scatter plot to determine the player represented by each point. Do any points appear to be outliers? Explain.

1996 Chicago Bulls

PLAYER	MINUTES	POINTS
Jordan	3090	2491
Pippen	2825	1496
Kukoc	2103	1065
Longley	1641	564
Kerr	1919	688
Harper	1886	594
Rodman	2088	351
Wennington	1065	376
Salley	673	185
Buechler	740	278
Simpkins	685	216
Brown	671	185

Table 8.2 (Source: Championship Commemorative Chicago Bulls Beckett Tribute)

3. Compare the 1994 Chicago Bulls to the 1996 Chicago Bulls by displaying both scatter plots simultaneously. What conclusions can you make? (To help discern between the scatter plots, use a different mark for each.)

Calculating and Graphing Least-Squares Regression Lines

Using the TI-83, you can calculate and graph least-squares regression lines for paired sets of data. For example, to calculate and graph a least-squares regression line for the data in lists MINS and PTS, do the following. (Turn off all stat plots except the scatter plot that represents the 1994 Chicago Bulls minute and point totals.)

> You can use the **LinReg(ax+b)** command to determine the least-squares regression line. Most statistics books, however, use the Greek letter beta, β, to represent slope. Therefore, **LinReg(a+bx)** is more appropriate.

1. Access the TI 83's **LinReg(a+bx)** command by pressing [STAT] [▶] [8]. The command will appear on the home screen.

2. Specify the lists to be used to determine the regression line by pasting the list names after **LinReg(a+bx).** Separate the list names using a comma. If no lists are specified, the default lists are L1 and L2.

3. Specify the Y= variable in which the equation of the regression line is to be stored. For example, to store the regression line equation in Y1, press [,] [VARS] [▶] [1] [1]. The home screen should display the following command.

   LinReg(a+bx) ʟMINS,ʟPTS,Y1

4. Press [ENTER]. The values *a* and *b* are displayed on the home screen and the equation of the least-squares regression line is stored in Y1.

> If you do not specify a Y= variable, the regression line equation is stored in **RegEQ**. You can access **RegEQ** by pressing [VARS] [5] [▶] [▶] [1].

5. Press [GRAPH]. The calculator display should be similar to the one shown below.

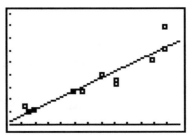

Classroom Exercises

1. Does the regression line appear to be a good fit? Explain.

2. Identify the slope and *y*-intercept of the regression line. What information does each provide?

3. Use **2-Var Stats** to determine the means of lists MINS and PTS. Then verify that the regression line contains the ordered pair (mean minutes, mean points).

Making Predictions Using a Regression Line

A primary reason for calculating a regression line is to obtain a model that can be used to make predictions. The TI-83 provides several methods for making predictions using a regression line. Each method is described below.

Making Predictions From the Home Screen

To make a prediction from the home screen, simply evaluate the regression equation using the explanatory value. For example, to predict how many points a player on the 1994 Chicago Bulls would have scored had he played 1800 minutes, enter the following keystrokes. (Assume the correct regression equation is stored in Y1.)

[VARS] [▶] [1] [1]	*Specify* Y= *variable.*
[(] 1800 [)]	*Specify value.*
[ENTER]	

The result is 750.11. Therefore, you can predict that if a player on the 1994 Chicago Bulls had played 1800 minutes, he should have scored approximately 750 points.

Making Predictions From the Graph Screen

To make a prediction from the graph screen, the regression line must be displayed. There are two ways to make a prediction directly from the graph screen: using the **value** command and using the trace feature.

To use the **value** command to predict how many points a player would have scored had he played 1800 minutes, enter the following keystrokes.

[2nd] [CALC] [1]	*Select* **value** *command.*
1800 [ENTER]	*Specify value.*

The resulting y-value, 750.11399, is displayed in the lower right corner of the viewing window.

To use the trace feature to predict how many points a player would have scored had he played 1800 minutes, press [TRACE] and use the up or down arrows to position the trace cursor on the regression line. Then enter 1800 and press [ENTER]. The resulting y-value, 750.11399, is displayed in the lower right corner of the viewing window.

Classroom Exercise

1. Suppose a player on the 1994 Chicago Bulls had played 1000 minutes. How many points should he have scored during the season?

Examining Residuals

Residuals provide an indication of how well a regression line models data. Whenever you use the TI-83 to calculate a regression model, the calculator also automatically calculates the residuals and stores them in a list named RESID. List RESID changes each time a regression analysis is performed. If you want to save a list of residuals, you must copy the contents of RESID into a new list before performing a subsequent regression analysis.

> List RESID is a "restricted" list. For example, you cannot use **1-Var Stats** to analyze list RESID. You can, however, use RESID when defining residual plots. Be very careful when using list RESID to examine data. For example, if you sort corresponding lists without sorting RESID or if you perform other regression analyses, list RESID will not contain the data you expect.

Classroom Exercises

1. From the stat list editor, create a list named BLRES. Then copy into list BLRES the residuals that result from calculating the 1994 Chicago Bulls regression line. Examine and discuss list BLRES.

2. Execute the following command from the home screen.

    SortA(LMINS,LPTS,LBLRES)

    How does executing this command affect the data in lists MINS, PTS, and BLRES? Examine list BLRES to observe that the sign of the residuals is fairly random.

3. Regardless of how well a regression line models a set of data, the sum of the residuals must be zero. Verify that this is true for the residuals in list BLRES.

4. Residuals are defined as the vertical deviations of the data points from the least-squares regression line. That is,

    residual = observed value − predicted value.

    Explain how to use the TI-83's list features to verify the residuals in list BLRES. Then verify the residuals in list BLRES.

Creating a Residual Plot

Viewing a residual plot often provides more meaningful information than simply examining a list of residuals. Creating a residual plot using the TI-83 is very much like creating a scatter plot. To create a residual plot, simply define a scatter plot and plot the residuals with respect to the original explanatory variable. For example, to create a residual plot of the data in lists MINS and BLRES, do the following.

1. Turn off any existing plots by entering

    [2nd] [STAT PLOT] [4] [ENTER].

2. Access Plot3 (stat plot editor) and define the residual plot.

    Begin defining the residual plot by selecting On and the scatter plot icon. (You can select each by moving the cursor over the appropriate item and pressing [ENTER].) Then paste list names MINS and BLRES after Xlist and Ylist, respectively. Finally, choose a data mark. After defining the scatter plot, the calculator display should look similar to the one shown below.

3. Define an appropriate viewing window in which to display the residual plot.

    You can define a viewing window by pressing [WINDOW] and entering appropriate values. It is much easier, however, to use the TI-83's **ZoomStat** command. **ZoomStat** automatically defines the viewing window settings and graphs the residual plot. To define a viewing window and graph the residual plot, enter [ZOOM] [9]. The residual plot should look similar to the one shown below.

    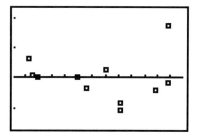

Linear Bivariate Relationships   43

The residual plot indicates no noticeable pattern regarding the occurrence of positive and negative residuals.

Classroom Exercises

1. Trace the residual plot. What do the positive residuals represent? The negative residuals? Identify the player associated with the greatest positive residual. Identify the player associated with the negative residual greatest in magnitude.

2. The residual plot indicates that the residuals do not occur in a noticeable pattern. What can you conclude from this result? Explain your reasoning.

Calculating Correlation Coefficients

Whenever the TI-83 calculates a regression line, it also calculates a corresponding correlation coefficient, r. You can obtain the correlation coefficient by accessing the r-value stored in the calculator or by instructing the calculator to display r automatically each time a regression analysis is performed.

To access the correlation coefficient stored in the TI-83's memory, first execute the command **LinReg(a+bx)** L**MINS,**L**PTS** from the home screen. This will restore the statistical variables to the TI-83's memory. Then enter

[VARS] [5] [▶] [▶] [7] [ENTER].

The result, $r \approx .9372$ is displayed on the home screen.

To display the correlation coefficient, r, automatically each time you perform a regression analysis, execute the **DiagnosticOn** command. To execute this command, enter the following keystrokes.

[2nd] [CATALOG]	*Access catalog menu.*
[D] Cursor to DiagnosticOn.	*Access command.*
[ENTER] [ENTER]	*Execute command.*

> Execute **DiagnosticOff** to turn off the automatic display.

Verify that r is displayed automatically by executing **LinReg(a+bx)** L**MINS,**L**PTS** again. Notice that the coefficient of determination, r^2, is also displayed.

Classroom Exercises

The following exercises extend or integrate the topics discussed in this chapter.

1. The formula for determining the correlation coefficient, r, is

$$r = \frac{\sum_{i=1}^{n}\left(\frac{x_i - \bar{x}}{Sx}\right)\left(\frac{y_i - \bar{y}}{Sy}\right)}{n - 1}$$

   Explain how to use this formula and the TI-83's list features to verify the correlation coefficient obtained for the 1994 Chicago Bulls minutes and point totals model. Then verify r.

2. Verify that the slope, b, of the 1994 Chicago Bulls regression line is

$$b = r\frac{Sy}{Sx}.$$

3. Verify that the y-intercept, a, of the 1994 Chicago Bulls regression line is

$$a = \bar{y} - b\bar{x}.$$

4. How does using a slightly different linear equation to model the 1994 Chicago Bulls minute and point totals affect any predictions and residuals? Investigate by changing the slope, the y-intercept, then both the slope and y-intercept. Discuss your results. Include any relevant summary statistics, residual plots, etc., in your discussion.

5. How does adding a constant to the dependent variable change the regression line equation and the corresponding correlation coefficient? Investigate by adding 50 points to the point totals of each player on the 1994 Chicago Bulls. How does the slope of the least-squares regression line change? The y-intercept? The correlation coefficient? Discuss your results.

6. How does increasing the dependent variable by a certain percent change the regression line equation and the corresponding correlation coefficient? Investigate by increasing the total points of each player on the 1994 Chicago Bulls by 10%. How does the slope of the least-squares regression line change? The y-intercept? The correlation coefficient? Discuss your results.

7. It can be argued that playing time is determined by the number of points a player scores. Using the data from the 1994 Chicago Bulls, create a scatter plot that represents this situation. Then calculate a least-squares regression line to model the data and find the corresponding correlation coefficient. How do your results compare to the original equation and correlation coefficient? Create a residual plot using the new results and compare it to the original. Discuss your results.

8. Calculate a least-squares regression line that models the 1996 Chicago Bulls minute and point totals. Does the model appear to be a good fit? Explain. Compare the resulting equation to the one obtained for the 1994 Chicago Bulls. What accounts for the differences between the models? Compare the correlation coefficients, residuals, and residual plots for each model. Which model accounts for the larger fraction of the variation in the y-variable? Explain.

9. Create a single scatter plot that represents the playing time and point totals for the 1994 *and* 1996 Chicago Bulls. Perform a regression analysis on these data to determine the equation of the least-squares regression line and the correlation coefficient. Examine the residuals. Discuss your results. Then compare the results to those obtained for the individual teams.

10. Does a correlation coefficient close to 0 always imply no relationship between two variables? Investigate by using the following data and calculating a least-squares regression line. Discuss your results and explain why the correlation coefficient is 0. Does a relationship exist between these lists? Explain.

    L1 = {−4, −3, −2, −1, 0, 1, 2, 3, 4}
    L2 = {16, 9, 4, 1, 0, 1, 4, 9, 16}

11. Does a correlation coefficient close to 1 always imply a strong linear relationship between two variables? Investigate by using the following data, calculating a least-squares regression line, and examining the residuals. Discuss the pattern of the residuals. Why is the correlation coefficient misleading?

    L1 = {1, 2, 3, 4, 5, 6, 7, 8, 9, 10}
    L2 = {1, 1.4, 1.7, 2, 2.2, 2.4, 2.6, 2.8, 3, 3.2}

You can combine two lists using the **augment** command from the list ops menu. For example, to combine lists MINS and NMINS and store the results in L4, execute the command **augment(LMINS, LNMINS)→L4.**

Investigating the Sensitivity of Regression and Correlation Calculations

The TI-83 allows students to investigate the effects of outliers, influential points, and other special data characteristics on regression and correlation calculations. The following sections outline two such activities.

Investigating the Effects of Influential Points

1. The following table lists the test scores for sixteen students in an introductory statistics class. Enter the Test 1 scores into L1 and the Test 2 scores in L2.

Student	Test 1	Test 2
1	96	92
2	100	96
3	88	76
4	83	82
5	95	95
6	95	89
7	85	69
8	97	92
9	63	54
10	83	71
11	81	84
12	92	92
13	93	94
14	95	99
15	72	69
16	91	86

Table 8.3

2. Create a scatter plot of the data in L1 and L2. Use the data in L1 as the explanatory variable. Discuss the scatter plot.

3. Perform a regression analysis to determine the equation of the least-squares regression line that models the data. Discuss the model's slope and *y*-intercept. What can you determine from the slope and *y*-intercept? Does the model appear to be a good fit? What is the correlation coefficient? What fraction of the variation in grades on the second test can be accounted for by the least-squares regression on the grades from the first test?

4. Delete the data point associated with the student having the lowest grades. Then create a scatter plot representing the revised data.

5. Find a least-squares regression line through the revised data. Discuss the slope, y-intercept, and correlation coefficient of the regression equation. How do the slope and y-intercept compare to the previous results? The correlation coefficient? Explain.

Investigating the Effects of Data Clusters

A basketball fan conjectures that personal fouls are highly correlated with turnovers. The fan makes this conjecture based on the belief that a player who loses possession of the basketball is likely to commit a foul while attempting to regain possession. In this activity, you will investigate this conjecture using the data in Table 8.4. The data provide the name and corresponding number of turnovers and personal fouls for each member of the 1996 University of North Carolina men's basketball team.

1996 University of North Carolina Men's Basketball

PLAYER	TURNOVERS	PERSONAL FOULS
Calabria	70	61
Carter	36	57
Geth	6	8
Jamison	59	69
Lynn	3	5
McInnis	78	72
McNairy	4	5
Neal	2	4
Okulaja	63	72
Sullivan	2	0
Tyndall	0	0
Williams	57	44
Zwikker	40	74

Table 8.4   (Source: University of North Carolina Sports Information)

1. Without performing any statistical analysis, do the data appear to support the fan's conjecture? Explain.

2. Enter the data into lists named UNCTO and UNCPF. Then calculate the equation of the least-squares regression line using the data in UNCTO as the explanatory variable. What is the correlation coefficient? What percent of the variation in personal foul totals can be accounted for by the regression of total fouls on total turnovers? Do the data appear to support the fan's conjecture? Explain.

3. Another basketball fan has a concern regarding the correlation coefficient obtained from the regression analysis. This fan believes that the high correlation coefficient results from a *lurking variable*, playing time. To determine whether playing time may be a lurking variable, create a scatter plot to represent the data in lists UNCTO and UNCPF. Discuss the scatter plot.

4. From the scatter plot, you can see that the data form two clusters separated by a large gap. One group consists of the 7 players who played the most during the season while the other consists of the 6 players who substituted occasionally. With respect to the cluster of points representing the players with the most playing time, do turnovers and personal fouls appear highly correlated? Do the data appear to support the original conjecture or the second fan's conjecture? Explain.

5. Create a scatter plot that represents the data for the seven players with the most playing time. To create the scatter plot, you can manually edit lists UNCTO and UNCPF to separate the data for the seven players from each list or you can use the TI-83's **Select** command. To separate the data using **Select,** first display the original scatter plot. Then press

    [2nd] [LIST] [▶] [8] [2nd] [L1] [,] [2nd] [L2] [)] [ENTER]

    to access the **Select** command and define the lists to which the selected data are to be stored. After pressing [ENTER], you are prompted to define a left bound. Define the left bound by positioning the trace cursor on the left-most desired data point and pressing [ENTER]. Define the right bound in a similar manner. After defining the bounds, notice that the scatter plot changes to represent the desired data. Do the data seem to indicate that total turnovers and total personal fouls are highly correlated? Explain.

6. Calculate the equation of the least-squares regression line using the data in L1 as the explanatory variable. What is the correlation coefficient? What can you conclude?

9 Transformations to Achieve Linearity

Not all bivariate data is best modeled by a linear equation. When a scatter plot or residual plot indicates that a nonlinear model is more appropriate, it is often helpful to transform the data to achieve linearity and find a linear model for the transformed data. Two such transformations are log-log and semi-log transformations. Each transformation has the effect of "straightening" certain nonlinear data.

In this chapter, you will learn how to use the TI-83 to perform each transformation. You will also learn how to use the TI-83 to perform power and exponential regressions.

Performing Log-Log Transformations

A log-log transformation requires taking the logarithm of the explanatory and response data values. A log-log transformation has the effect of straightening data that can be modeled by a power function of the form $y = ax^n$.

The following data list several lengths, in centimeters, and corresponding periods, in seconds, of a swinging pendulum.

Swinging Pendulums

LENGTH	PENDULUMS
6.5	0.51
18.0	0.88
30.5	1.13
41.5	1.27
53.2	1.45
65.1	1.64
80.0	1.77

Table 9.1

To perform a log-log transformation and find a model for the pendulum data, do the following.

1. Use **SetUpEditor** to create two lists named LEN and PER. Then store the pendulum lengths and periods in lists LEN and PER, respectively. Using the data in list LEN as the explanatory values, create a scatter plot to represent the pendulum data. What can you conclude from the scatter plot? Perform a linear regression analysis and use the results to create a residual plot. What can you conclude?

2. Create and display list names LOGX and LOGY in the stat list editor. These lists will contain the transformed data. To create and display list name LOGX, access the stat list editor, cursor to the empty column heading, and type the list name. Create and display list name LOGY in a similar manner. After creating list names LOGX and LOGY, define each as shown below.

    $\text{LOGX} = \log(\text{LLEN})$ $\qquad$ $\text{LOGY} = \log(\text{LPER})$

3. Using the data in list LOGX as the explanatory values, create a scatter plot to represent the data in lists LOGX and LOGY. (Turn off the scatter plot created in Part 1.) Discuss the scatter plot. How does it compare to the scatter plot created in Part 1? Do the data appear to be "straightened?" Is a power function an appropriate model for the pendulum data? Explain.

4. Perform a linear regression on the data in lists LOGX and LOGY. Store the regression equation in Y1. Graph the regression line and scatter plot simultaneously. Does the regression line appear to be a good fit? Identify the slope and y-intercept of the regression line. What is the correlation coefficient? What can you conclude from the correlation coefficient?

5. Confirm that the regression line is a good fit by examining the residuals and creating a residual plot. How does the residual plot compare to the one obtained in Part 1?

6. Now that you have determined the relationship between the transformed data, you can transform the regression line to find a model for the original data. Using the following algebraic techniques and the model found in Part 4, find a power function that models the original data.

$$\log(y) = a + b\log(x)$$
$$10^{\log(y)} = 10^{(a + b\log(x))}$$
$$10^{\log(y)} = (10^a)(10^{b\log(x)})$$
$$10^{\log(y)} = (10^a)(10^{\log(x^b)})$$
$$y = (10^a)x^b$$

What is the relationship between the linear equation and the power in the power function?

When entering the power function into the Y= editor, use TI-83's **Rcl** command to recall the exact values of *a* and *b*. To enter the equation using **Rcl,** access an empty Y= variable and enter

10 [^] [2nd] [RCL] [VARS] [5]
[▶] [▶] [2] [ENTER] [X,T,Θ,n]
[^] [2nd] [RCL] [VARS] [5] [▶]
[▶] [3] [ENTER].

The TI-83 is programmed to perform a power regression analysis using a least-squares fit and the transformed values ($\ln(x)$, $\ln(y)$). It is important to realize that the values r and r^2 represent the strength of the linear relationship between the transformed data values.

7. Graph the power function obtained above and the original scatter plot on the same viewing screen. (Turn off any other scatter plots and Y= variables.) Is the model a good fit?

8. Use the power function to obtain a list of residuals. Store the residuals in a list named PENRS and use the list to create a residual plot. Discuss the residual plot. Are there any noticeable patterns? How can you tell from the residual plot that the power function is a good fit? How does this residual plot compare to the one in Part 1?

9. Use the model to predict the period of a pendulum having a length of 35 cm. Predict the period of a pendulum having a length of 100 cm.

10. Using the TI-83's **PwrReg** command, you can directly perform a power regression on a set of paired data. For example, to perform a power regression on the data in lists LEN and PER, enter the following keystrokes.

[STAT] [▶] [ALPHA] [A]   *Access* **PwrReg** *command.*
[2nd] [LIST] Choose LEN.   *Choose appropriate lists.*
[,] [2nd] [LIST] Choose PER.
[ENTER]

Discuss the relationship between the resulting power function and the model obtained in Part 6.

Performing Semi-Log Transformations

A semi-log transformation requires taking the logarithm of the response data values while leaving the explanatory data values unchanged. A semi-log transformation has the effect of straightening data that can be modeled by an exponential function of the form $y = ab^x$.

Population counts with respect to time can often be modeled by an exponential function. The following table lists the combined American Indian, Eskimo, and Aleut population (in thousands) for selected years.

American Indian, Eskimo, and Aleut Population

YEAR	POPULATION
1950	377
1960	552
1970	827
1980	1420
1990	1959

Table 9.2   (Source: U.S. Bureau of the Census)

To perform a semi-log transformation and find a model for the population data, do the following.

1. Use **SetUpEditor** to create two lists named YR and POP. Then store the census data into these lists. Using the data in list YR as the explanatory values, create a scatter plot of the census data. What can you conclude from the scatter plot?

2. Create and display a list named LOGP in the stat list editor. Define LOGP = log(LPOP). Then using the data in list YR as the explanatory values, create a scatter plot to represent the data in lists YR and LOGP. (Turn off any existing plots and Y= variables.) Discuss the scatter plot. Do the data appear to be "straightened"? Is an exponential equation an appropriate model for the census data? Explain.

3. Perform a linear regression on the data in lists YR and LOGP. Store the regression equation in Y1. Graph the regression line and the scatter plot simultaneously. Does the regression line appear to be a good fit? Identify the slope and y-intercept of the regression line. What is the correlation coefficient? What can you conclude from the correlation coefficient?

4. Confirm that the regression line is a good fit by examining the residuals and creating a residual plot. Are there any noticeable patterns?

5. Now that you have determined the linear relationship between the year and the transformed population data, you can find an exponential model for the original data. Using the following algebraic techniques and the linear model found in Part 3, find an exponential equation to model the original data.

$$\log(y) = a + bx$$
$$10^{\log(y)} = 10^{(a+bx)}$$
$$y = (10^a)(10^b)^x$$

What is the relationship between the constants in the exponential equation and the constants in the linear equation?

Use the TI-83's **Rcl** command to recall the exact values of *a* and *b* when entering the exponential equation into a Y= variable.

6. Graph the exponential equation and the original scatter plot simultaneously. (Turn off any existing plots and Y= variables.) Is the model a good fit? Explain.

7. Use the exponential equation to obtain a list of residuals. Store the residuals in a list named CPRES and use the list to create a residual plot. Discuss the residual plot. Are there any noticeable patterns? How can you tell from the residual plot that the exponential model is a good fit?

> The TI-83 is programmed to perform an exponential regression analysis using a least-squares fit and the values $(x, \ln(y))$. It is important to realize that the r value represents the strength of the linear relationship between the explanatory values and the transformed response values. The r^2 value represents the fraction of the variation in transformed response values accounted for by the linear regression on the explanatory values.

8. Using the TI-83's **ExpReg** command, you can directly perform an exponential regression on the data in lists YR and POP. To perform an exponential regression on the data in these lists, enter the following keystrokes.

   STAT ▶ 0 *Access **ExpReg** command.*
   2nd [LIST] Choose YR. *Choose appropriate lists.*
   , 2nd [LIST] Choose POP.
   ENTER

   Discuss the relationship between the resulting exponential equation and the model obtained in Part 5.

9. Predict the combined American Indian, Eskimo, and Aleut population for the year 2000. How confident are you with respect to this prediction? Explain.

Classroom Exercises

1. **Kepler's Third Law**   In 1618 Johann Kepler discovered the mathematical relationship between the distance from the sun to each planet of our solar system and the time required for the planet to complete one revolution of the sun. This relationship, known as Kepler's Third Law, can be calculated by transforming the data in Table 9.3 and then fitting a least-squares regression line to the linearized data. The data in Table 9.3 list each planet's mean distance from the sun (in millions of miles) and its period of revolution. Use the transformation techniques discussed in this chapter to "straighten" the data. Then find a linear equation that models the linearized data and use the equation to determine a model for the original data. Determine the validity of the model by graphing the model and its corresponding scatter plot on the same viewing screen. Also examine the residuals and graph a residual plot. Discuss your results.

Planetary Statistics

PLANET	MEAN DISTANCE FROM SUN	PERIOD OF REVOLUTION
Mercury	36.0	0.241
Venus	67.0	0.615
Earth	93.0	1.000
Mars	141.5	1.880
Jupiter	483.0	11.900
Saturn	886.0	29.500
Uranus	1782.0	84.000
Neptune	2793.0	165.000
Pluto	3670.0	248.000

Table 9.3

2. Use the TI-83 to find a model for the data in Table 9.3 (without transformation) and to verify the model found in Exercise 1. Interpret the values of r and r^2 provided with this calculation.

Additional Curve Fitting Tips

1. Because the TI-83's power regression analysis is based on a log-log transformation, all **explanatory** and **response** values must be positive to avoid an error message.

2. Because the TI-83's exponential regression analysis is based on a semi-log transformation, all **response** values must be positive to avoid an error message.

10 Using Simulations to Estimate Probabilities

If this is the first time that your students have used the TI-83's random number generator features, instruct each student to store a different seed value to **rand** before starting these activities. If necessary, refer to page 22 for more information on storing a seed value.

This chapter contains two examples that demonstrate how to estimate probabilities using simulations performed by the TI-83. The simulations are performed using the TI-83's random integer generator, **randInt,** and random binomial generator, **randBin.**

Simulating Baseball Probabilities

The TI-83's random binomial generator, **randBin,** generates and displays a random number from a specified binomial distribution. The generated number represents the number of successes in a binomial experiment for a specified number of trials and probability of success. Consider the following problem, which develops a simulation using **randBin.**

> In the Major League Baseball World Series, the American and National League teams play each other until one of the teams wins four games. Suppose the team from the National League is considered the better team with a probability of 0.65 to win any individual game. What is the probability that the World Series will last more than five games?

To estimate the probability that the World Series will last more than five games, you can simulate the results of five World Series games, tally the results for a number of series, and calculate the empirical probability.

To simulate the results of five World Series games, enter the following keystrokes.

MATH ◄ 7   *Access* **randBin** *command.*
1 , .65 , 5)   *Specify binomial distribution.*
ENTER

After entering the keystrokes, you should obtain a list containing five elements. Each element is a 0 or a 1. The probability of a 1 is 0.65, so let 1 represent a win by the National League team. If the resulting list contains four or more 0's or 1's, you can conclude that the series does not extend beyond five games.

To estimate the probability that the World Series will last more than five games, simply repeat the simulation a number of times, tally the results, and then use the results to calculate the estimated probability. To perform the simulation again, simply press ENTER.

Classroom Exercises

1. Perform the simulation fifty times to simulate the results of fifty World Series. Estimate the probability that the World Series extends beyond five games.

2. Calculate the theoretical probability that the World Series extends beyond five games.

3. To simulate the results of one hundred World Series, execute the command

    randBin(5, .65,100) → L1

    from the home screen. Executing this command instructs the calculator to perform one hundred simulations to determine how many games out of five the National League wins. The results of each simulation—0, 1, 2, 3, 4, or 5—are stored in L1. What results indicate that the National League team won in five or fewer games? What results indicate that the American League team won in five or fewer games? What results indicate that the World Series extends beyond five games?

4. Create a histogram to display the results of Exercise 3. Let Xmin = −.5, Xmax = 5.5, and Xscl = 1. Trace the histogram to determine the frequency of each outcome. Estimate the probability that the World Series extends beyond five games. Compare your results with those obtained in Exercises 1 and 2.

Simulating the Birthday Problem

The TI-83's random number generator, **randInt,** generates and displays a random integer within a specified range. Consider the "Birthday Problem"— a classic statistics problem which can be simulated using **randInt.**

> In a class of 25 students, what is the probability that at least two students share the same birthday (month and day only, not necessarily year)?

To simulate the birthdays of 25 students, enter the following keystrokes.

Keystrokes	Description
[MATH] [◄] [5]	*Access* **randInt** *command.*
1 [,] 365 [,] 25 [)]	*Specify range and number of trials.*
[STO▶] [2nd] [L1]	*Store results.*
[ENTER]	

After entering the keystrokes, you should obtain 25 entries in list L1. Each entry is a number from 1 to 365. The entries represent the days of the year. For example, the entry 10 represents January 10, the tenth day of the year. If two entries are equal, you can conclude that at least two students share the same birthday. Use the TI-83's **SortA** command to sort L1. Then cursor through L1 to determine whether there are any identical entries.

To estimate the probability that at least two students share the same birthday, repeat the simulation a number of times, tally the results, and calculate the empirical probability.

There are several ways to enhance this simulation. For example, to identify easily any identical entries in L1, enter the following keystrokes from the home screen immediately after sorting L1.

Keystrokes	Description
[2nd] [LIST] [▶] [7]	*Access ΔList command.*
[2nd] [L1] [)]	*Define list.*
[STO▶] [2nd] [L2]	*Store results.*
[ENTER]	
[STAT] [2] [2nd] [L2] [)]	*Sort results.*
[ENTER]	
[2nd] [L2] [ENTER]	*Display results*

Command **ΔList** calculates the difference between consecutive elements in a specified list. After entering the keystrokes given above, you should obtain a list in L2 that contains 24 entries. Each entry is the difference between entries in L1. A 0 entry implies that at least two students share the same birthday.

Another way to enhance the simulation is to combine all the instructions. This makes it easier to repeat the simulation and analyze the results. To perform a number of simulations quickly, execute the following commands from the home screen and press [ENTER] repeatedly. Be sure to separate the commands using a **:**.

randInt(1,365,25) → L1:SortA(L1):ΔList(L1) → L2:SortA(L2):L2

Each time you press [ENTER], L1 and L2 are updated and L2 is displayed on the home screen. If the display begins with a 0, you can conclude that at least two students share the same birthday.

Classroom Exercises

1. Perform the simulation fifty times and tally the results. Estimate the probability that at least two students in a class of 25 share the same birthday.

2. What is the theoretical probability that in a class of 25 students at least two students share the same birthday?

3. Modify the simulation to estimate the probability that at least two students in a class of 24 share the same birthday. Repeat for a class of 23 students. Repeat for a class of 22 students. If the probability that two students share the same birthday is 0.5, how many students are in the class?

4. Suppose that while performing the simulations above, you notice that L2 begins with two or more 0 entries. What do multiple 0 entries indicate? (*Hint*: There are several answers.)

5. Design a simulation to estimate the probability that three students in a class of 25 share the same birthday.

11 Binomial Probabilities

In this chapter, you will use the TI-83 to calculate and investigate binomial probabilities. Specifically, you will use the TI-83 to

- create a binomial probability distribution.
- create a binomial probability histogram.
- determine summary measures for a binomial distribution.
- approximate a binomial distribution.

Creating a Binomial Probability Distribution

The TI-83's list features enable you to store the probability distribution for a discrete random variable in tabular form. For example, consider tossing fifty coins and counting the number of resulting heads. To create a probability distribution for the binomial random variable, X, where X represents the number of heads, do the following.

1. Clear lists L1 and L2. Lists L1 and L2 will contain the values of X and the associated probabilities, respectively.

2. Enter the possible values of X into list L1. Because the number of heads can range from 0 to 50 inclusive, L1 must contain the integers 0 to 50. You can enter these integers individually; however, it is much more efficient to use the TI-83's **seq** command. To define L1 using **seq,** execute the following command from the home screen.

    seq(X,X,0,50,1) → **L1**

3. Determine the associated binomial probabilities and store the results in L2. There are two ways to use the TI-83 to calculate binomial probabilities. One way is to use the formula for determining binomial probabilities

$$P(X = k) = \binom{n}{k} p^k (1-p)^{n-k}.$$

Command **nCr** is found in MATH PRB menu.

For example, considering this situation where $n = 50$, $p = 0.5$ (a fair coin), and k represents each value in L1, you can calculate the binomial probabilities and store the results in L2 by executing the following command from the home screen.

(50 nCr L1)(.5^L1)(.5^(50-L1)) → **L2**

A more efficient way to calculate the probabilities is to use the TI-83's **binompdf** command. To use **binompdf** to calculate the probabilities and store the results in L2, enter the following keystrokes.

[2nd] [DISTR] [0]	*Access* **binompdf** *command.*
50 [,] .5 [,]	*Specify number of trials and probability of success.*
[2nd] [L1] [)]	*Specify number of successes.*
[STO▸] [2nd] [L2]	*Store probabilities.*
[ENTER]	

After entering the keystrokes, examine L2 to verify the probabilities. For example, the probability $P(X = 25)$ is approximately 0.1123.

Creating a Binomial Probability Histogram

To represent the binomial probability distribution (the coin toss data in L1 and L2) using a probability histogram, do the following.

1. Turn off any existing stat plots by entering

    [2nd] [STAT PLOT] [4] [ENTER].

2. Access Plot1 (stat plot editor) and define the histogram.

    Begin defining the histogram by selecting On and the histogram icon. (You can select each by moving the cursor over the appropriate item and pressing [ENTER].) Then define Xlist = L1 and Freq = L2. After defining the histogram, the calculator display should look identical to the one shown below.

3. Define an appropriate viewing window in which to display the histogram.

   Define the viewing window by pressing [WINDOW] and entering appropriate values. In this case, define Xmin = 12.5, Xmax = 37.5, Xscl = 1, Ymin = −.03, and Ymax = .12. Using these window settings centers the histogram's bars over the most likely values of X.

4. Press [GRAPH] to graph the histogram. The histogram should look similar to the one shown below.

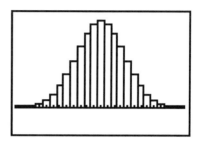

Determining Summary Measures for a Binomial Distribution

To determine the summary measures for a binomial distribution, use the TI-83's **1-Var Stats** command. For example, to determine the summary measures for the binomial distribution in L1 and L2, enter the following keystrokes.

[STAT] [▶] [1]	*Access* **1-Var Stats** *command.*
[2nd] [L1] [,]	*Specify X values.*
[2nd] [L2]	*Specify binomial probabilities.*
[ENTER]	

After pressing [ENTER], notice that the mean is $np = (50)(.5) = 25$ and the standard deviation is $\sqrt{np(1-p)} = \sqrt{50 \times 0.5 \times 0.5} \approx 3.5355$. Notice also that S_x is left blank as the sample standard deviation has no meaning in this context.

Binomial Probabilities 63

Approximating a Binomial Distribution

You can approximate the binomial distribution in L1 and L2 using a normal distribution. Because the approximating normal distribution has the same mean and standard deviation as the binomial distribution, define the approximating normal distribution to be $N(25, \sqrt{50 \times 0.5 \times 0.5})$. To show graphically that the normal curve can approximate the probability histogram for the binomial distribution in L1 and L2, do the following.

1. Access the Y= editor and turn off any existing Y= variables.

2. Cursor to an empty Y= variable and enter the following normal probability density function.

    normalpdf(X,25,$\sqrt{(50*.5*.5)}$)

3. Press [GRAPH] to display the probability histogram and the graph of the normal probability density function. The display should be similar to the one shown below.

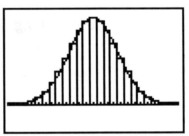

Entering

binompdf(50,.5,25)

from the home screen also yields an exact probability of obtaining 25 heads.

Trace the histogram until the trace cursor is on the bar having endpoints 24.5 and 25.5. The displayed *n*-value, 0.112275, represents the exact probability of obtaining 25 heads. To obtain the normal approximation for this probability, you must calculate the area under the normal curve between 24.5 and 25.5. To calculate this area, execute

normalcdf(24.5,25.5,25, $\sqrt{(50*.5*.5)}$)

from the home screen. The result is approximately 0.112463.

Classroom Exercises

1. Verify that the sum of the binomial probabilities in L2 is exactly 1.

2. When graphing the normal probability density function and the probability histogram, notice that the histogram's bars extend beyond the normal curve. Investigate how closely the normal curve approximates the probability histogram by changing the window settings and regraphing the histogram and normal curve. Discuss your results.

3. What is the probability of obtaining between 20 and 30 heads inclusive when tossing 50 coins? Use **binomcdf** to calculate the exact probability using the binomial distribution. Then approximate the probability using the normal approximation and **normalcdf.** Discuss your results.

4. Consider tossing 200 coins and counting the resulting number of heads. Create a probability distribution for the binomial random variable X, the number of heads. Represent the binomial distribution using a probability histogram. Then approximate the distribution with a normal probability density function. (*Note*: Define Xmin = 79.5, Xmax = 120.5, Ymin = −0.01, and Ymax = .06.)

5. Use the binomial distribution to calculate the exact probability of obtaining 100 heads when 200 coins are tossed. Then determine the probability of obtaining between 90 and 110 heads inclusive.

6. Use the normal approximation to estimate the probabilities in Exercise 5. Discuss your results. How do the calculated and approximated probabilities compare to the probabilities obtained for 50 coins? Why?

7. Suppose a teacher gives a multiple-choice exam consisting of 100 questions, each having four answer choices. Calculate the exact probability that you will get 60 or more questions correct by randomly guessing.

> The command **binomcdf(50,.5,30)** calculates the probability of obtaining 30 or fewer heads when 50 coins are tossed.

12 The Sampling Distribution of Sample Proportions

Using simulations, students can investigate, discover, and verify how sample statistics vary from one sample to another. In this chapter, you will learn how to use the TI-83 to perform simulations that enable students to investigate the sampling distribution of sample proportions. Specifically, you will use the TI-83 to

- investigate the variability of sample proportions.
- investigate the distribution of sample proportions.

Investigating the Variability of Sample Proportions

In this activity, you will perform simulations to investigate the relationship between sample size and the variability in sample proportions. Consider the following problem.

> Your friend challenges you to a friendly wager. If you toss a coin twenty times and get between 40% and 60% heads, you win. Otherwise, your friend wins. Before tossing the coin twenty times, two other friends offer some advice. One says that you have a better chance of winning if you toss the coin only ten times, while the other says your chances of winning increase when you toss the coin fifty times. How many coin tosses provide the best chance for you to win?

To easiest way to simulate coin tosses using the TI-83 is to use the **randBin** command. For example, to simulate twenty coin tosses, enter **randBin(20,.5)** from the home screen. The resulting display is the number of successes (heads).

To obtain the *proportion* of successes rather than the *number* of successes, divide the result by the number of tosses. For example, to simulate the proportion of heads for twenty coin tosses, enter **randBin(20,.5)/20** from the home screen. To perform fifty similar simulations, store the results in L1, and sort the results, enter the following commands from the home screen.

   randBin(20,.5,50)/20 → L1:SortA(L1)

After entering the above commands, examine list L1 to determine how many simulations resulted in 40% to 60% heads.

Classroom Exercises

1. Modify the simulation to toss a coin ten times and store the sample proportions in L2. How many simulations result in 40% to 60% heads?

2. Modify the simulation to toss a coin fifty times and store the sample proportions in L3. How many simulations result in 40% to 60% heads?

3. Based on your results, how many coin tosses, ten, twenty, or fifty, provide the best chance for you to win the friendly wager? Explain.

Investigating the Distribution of Sample Proportions

In this activity, you will perform a simulation to generate a list of sample proportions. You will then analyze the resulting distribution using summary statistics and graphs. Consider the following problem.

> The student body at a large university consists of 60% females and 40% males. Suppose you randomly select fifty students to participate in a survey regarding campus issues. What proportion of the fifty students can you expect to be female?

To perform 100 simulations and generate the desired list of sample proportions for a sample size of 50, execute the following command from the home screen.

randBin(50,.6,100)/50 → L1

The resulting entries in list L1 represent the proportion of fifty randomly selected students that are female.

To analyze the summary statistics associated with the distribution of sample proportions, use the TI-83's **1-Var Stats** command. The mean value of the sample proportions should be approximately 0.6. What is the minimum proportion value? The maximum proportion value? What is the standard deviation of the sample proportions?

To analyze the shape of the distribution of sample proportions, use a histogram. Before graphing the histogram, turn off any existing Y= variables and stat plots. Then in Plot1, define the histogram using Xlist = L1 and Freq = 1. When defining the window settings, Xmin should be less than or equal to the minimum proportion value, **min(L1),** and Xmax should be greater than or equal to the maximum proportion value, **max(L1).** Let Xscl = .02. Define Ymin and Ymax as necessary. After specifying the histogram and window settings, press [GRAPH].

The resulting histogram displays the shape of the distribution of sample proportions. Study the histogram. Are there any outliers? Most likely the histogram will appear reasonably symmetric and mound-shaped. Trace the histogram to the rectangle that contains the sample proportions equal to 0.6. Does this appear to be the approximate center of the distribution?

The histogram should appear somewhat normal. To determine whether the distribution is approximately normal, use the 68–95–99.7 Rule.

1. Use the summary statistics, $\bar{x}$ and Sx, to determine each interval. The first interval is $\bar{x} \pm Sx$; the second interval is $\bar{x} \pm 2Sx$; and the third interval is $\bar{x} \pm 3Sx$.
2. Sort L1 and count the number of observations in each interval.
3. Determine the percent of proportions in each interval.

Does the distribution satisfy the 68–95–99.7 Rule?

Classroom Exercises

1. Repeat the simulation described above to investigate the distribution of sample proportions for samples of size 100 and samples of size 200. Store the sample proportions in L2 (samples of size 100) and L3 (samples of size 200). What is the most obvious difference in the distributions? Record the standard deviation associated with each distribution. (*Note*: It will take the TI-83 several minutes to generate each list of sample proportions.)

2. Mathematical theory states that a distribution of sample proportions is approximately normal and that the distribution becomes more normal as the sample size increases. Are your simulations consistent with this theory? Explain.

3. Create modified box plots to represent the sample proportions in L1, L2, and L3. Graph the box plots on the same viewing screen. How do the medians compare? How do the ranges and interquartile ranges compare?

4. As the sample size increases, how does the standard deviation change? Investigate by performing 100 simulations for samples of size 25 and for samples of size 150 and recording the standard deviation for each. Then combine the results with previous sample sizes and standard deviations to create a scatter plot that represents standard deviation versus sample size. Perform a regression analysis to determine the relationship between sample size, n, and standard deviation, s. Discuss your results.

> Encourage students to perform the regression analysis by transforming the data using a log-log transformation. See Chapter 9.

5. Mathematical theory states that

   - The mean of a distribution of sample proportions equals the population proportion, p, from which the samples are drawn.

   - The standard deviation of a distribution of sample proportions is
   $$\sqrt{\frac{p(1-p)}{n}},$$
   where n is the sample size and p is the population proportion.

   Therefore, in each simulation performed in this activity, the mean should be approximately 0.6 and the standard deviation should be approximately
   $$\sqrt{\frac{(0.6)(0.4)}{n}} = \sqrt{\frac{0.24}{n}} = 0.49n^{(-0.5)}.$$
   Compare these theoretical relationships to the results of Exercise 4. Discuss your results.

6. Suppose 30% of the undergraduates at the university are freshmen. Use the results of Exercise 5 to describe the distribution of sample proportions of freshmen resulting from samples of size 100 taken from the undergraduate population. Then simulate fifty of these samples and determine how the experimental results compare to the theoretical results.

7. Graph a normal probability plot of the sample proportions created in Exercise 6. Do these sample proportions appear to be normally distributed? Explain.

13 The Sampling Distribution of Sample Means

In addition to using simulations to examine sample proportions, students can use simulations to investigate, discover, and verify how sample means vary from one sample to another.

In this chapter, you will use the TI-83 to perform simulations that enable students to investigate the sampling distribution of sample means. Specifically, you will use the TI-83 to

- investigate a distribution of sample means from a normal population.
- investigate a distribution of sample means from a population that is not normal.

Investigating a Distribution of Sample Means from a Normal Population

In this activity, you will use the TI-83 to simulate sample means, analyze a distribution of sample means, and investigate the normality of a distribution of sample means. Consider the following problem.

> A machine used to fill quart-sized milk containers is set at a target amount of 32.5 ounces. The standard deviation for this machine is 0.25 ounces. If you randomly choose four containers of milk and carefully measure the contents, what sample mean can you expect? How likely is it that the sample mean is less than 32.5 ounces? How likely is it that the sample mean is less than 32.25 ounces? What would be a surprisingly low mean or a surprisingly high mean?

Simulating Sample Means

There are two ways to use the TI-83 to simulate a sample mean drawn from a normal population. Each way uses the **randNorm** command. For example, you can use **randNorm** to simulate the amount of milk in each of the four containers and then combine the results to determine the sample mean. Using the normal distribution specified in the problem, you can simulate the amount of milk in one container by executing **randNorm(32.5, .25)** from the home screen. Repeat this command three times to simulate the amount of milk in four containers. What is the sample mean?

A more efficient method for obtaining the sample mean amount of milk in four milk containers is to perform a simulation that samples four containers at once and then use the **mean** command to find the mean of the four sample contents. You can accomplish this by executing the following command from the home screen.

mean(randNorm (32.5, .25, 4))

Execute the command twenty times. Observe that the sample means tend to be close to 32.5 and that very few sample means are below 32.25 or above 32.75. To determine how the means are distributed, it is necessary to perform many simulations. Using the TI-83's **seq** command, you can perform many simulations and create a list of sample means. To create one hundred sample means and store the results in list L1, enter the following command from the home screen.

seq(mean(randNorm(32.5,.25,4)),X,1,100,1) → L1

After executing the command, examine list L1 and verify that it contains 100 entries. Each entry represents the mean from a sample of size four. Each sample is drawn from the normal population $N(32.5, 0.25)$.

Analyzing a Distribution of Sample Means

To analyze a distribution of sample means, use the TI-83 to

- determine the summary statistics associated with the distribution.
- graph the distribution.

To determine the summary statistics associated with the distribution of sample means, execute **1-Var Stats L1** from the home screen. The mean of the sample means should be approximately equal to 32.5. What is the standard deviation of the sample means? What is the minimum sample mean? Maximum sample mean?

To graph the distribution, define Plot1 to be a histogram with Xlist = L1 and Freq = 1. Use **ZoomStat** to define the window settings and graph the histogram. The histogram should be fairly symmetric and mound-shaped. Trace the histogram to the bar that contains the sample means equal to 32.5. Does this appear to be the center of the distribution? What conclusions can you make by observing the histogram?

Investigating the Normality of a Distribution of Sample Means

The histogram representing the distribution of sample means should appear approximately normal. To determine whether the distribution is approximately normal, use the 68–95–99.7 Rule as outlined below.

1. Use the summary statistics associated with the distribution to determine each interval. The first interval is $\bar{x} \pm Sx$; the second interval is $\bar{x} \pm 2Sx$; the third interval is $\bar{x} \pm 3Sx$.

2. Sort L1 and count the number of observations in each interval.

3. Determine the percent of means in each interval.

Does the distribution of sample means satisfy the 68–95–99.7 Rule? Is it reasonable to conclude that the sample means are normally distributed? Explain.

Classroom Exercises

1. Repeat the simulation described above to investigate the distribution of sample means for samples of size 25 and for samples of size 50. Store the sample means in lists L2 (samples of size 25) and L3 (samples of size 50). What is the most obvious difference in the distribution of these means and the distribution generated for samples of size 4? Record the standard deviation associated with each distribution. (*Note*: It will take the TI-83 several minutes to generate each distribution of sample means.)

2. Mathematical theory states that distributions of sample means drawn from normally distributed populations are normally distributed, even when sample sizes are very small. Use normal probability plots to determine whether your simulations are consistent with this theory.

3. Create modified box plots to represent the sample means in L1, L2, and L3. Graph the box plots on the same viewing screen. How do the medians compare? How do the ranges and interquartile ranges compare? What conclusions can you make? Explain.

4. As the sample size increases, how does the standard deviation of the distribution of sample means change? Investigate by performing 100 simulations for samples of size 10 and for samples of size 100. Record the standard deviation for each. Then combine the results with previous sample sizes and standard deviations to create a scatter plot that represents standard deviation versus sample size. Perform a regression analysis to determine the relationship between sample size, *n*, and standard deviation, *s*. Discuss your results.

> Encourage students to perform the regression analysis using a log-log transformation. See Chapter 9, if necessary.

5. Mathematical theory states that

    - The mean of a sampling distribution of sample means equals the population mean, μ, from which the samples are drawn.

    - The standard deviation of a sampling distribution is
    $$\frac{\sigma}{\sqrt{n}},$$
    where σ is the population standard deviation and n is the sample size.

    Therefore, in each simulation performed in this activity, the mean should be approximately 32.5 and the standard deviation should be approximately $0.25/\sqrt{n}$. Compare these theoretical relationships to the results of Exercise 4. Discuss your results.

Investigating a Distribution of Sample Means from a Non-Normal Population

Suppose you select simple random samples from populations that are not normally distributed and calculate the sample means. How will the mean and standard deviation of these sample means compare to the mean and standard deviation of the underlying population? Will the distribution of sample means be normally distributed? In this activity, you will investigate a sampling distribution of sample means from a population that is not normal by using the TI-83 to select random samples from the population, calculating sample means, and analyzing the distribution of sample means.

In the activity, you will use the data listed in Table 1.1 on page 1. If list REIGN is stored in your TI-83, insert it in the stat list editor. Otherwise, enter the data. The data in this list are not normally distributed. The distribution is skewed to the right. Before beginning the activity, you may want to define and examine a histogram of the data in list REIGN to verify that the underlying population from which you draw your samples is not normally distributed. Also, use **1-Var Stats** to verify that $\mu = 21.875$ and $\sigma = 16.219$. Note that list REIGN contains forty entries from which you will draw samples of size four.

Obtaining Sample Means

To obtain a sample mean from the data in list REIGN, execute the command

mean(seq(LREIGN(randInt(1,40)),X,1,4,1))

from the home screen. This command instructs the TI-83 to create a list of four randomly selected entries from list REIGN and to calculate the mean of the four entries. Execute the command several times. You should notice variation in the results. To determine how the means are distributed create a list consisting of many sample means from the population of reign lengths.

To create a list of sample means from the population of monarch reigns and store the results in list L2, do the following.

1. Clear list L2 by pressing [STAT] [4] [2nd] [L2] [ENTER].

2. Define C to be a counter variable with an initial value of 1 by entering

    1 [STO▶] [ALPHA] [C].

3. Execute the following command from the home screen.

    mean(seq(LREIGN(randInt(1,40)),X,1,4,1)) → L2(C):C+1 → C

4. After executing the command in Part 3 above, you will see the number 2 displayed on the home screen. This indicates the value of C and that the calculator is waiting to calculate another sample mean. Press [ENTER] repeatedly until the calculator displays the number 101 on the home screen.

Examine list L2 and verify that it contains 100 entries. Each entry represents the mean from a sample of size four. Each sample is drawn from list REIGN.

Analyzing a Distribution of Sample Means

To analyze a distribution of sample means, use the TI-83 to

- determine the summary statistics associated with the distribution.
- graph the distribution.

To determine the summary statistics associated with the distribution of sample means, execute **1-Var Stats L2** from the home screen. The mean of the sample means should be approximately equal to 21.875. The standard deviation should be approximately 8.110. What is the minimum sample mean value? Maximum sample mean value? What can you conclude from these sample mean values?

To graph the distribution, define Plot1 to be a histogram with Xlist = L2 and Freq = 1. Use **ZoomStat** to define the window settings and graph the histogram. Describe the histogram's shape. What conclusions can you make from the histogram? Another way to graph the distribution is to define Plot1 to be a normal probability plot with DataList = L2. Use **ZoomStat** to define the window settings and graph the plot. What conclusions can you make from the normal probability plot?

Statistical theory states that the mean and standard deviation of sample means resulting from simple random samples drawn from any population will conform to the rules stated in Exercise 5 on page 74 for samples drawn from normal populations. That is, the sampling distribution of sample means will always have the same mean as the population, μ, and the standard distribution of the sample means will be $\sigma/\sqrt{n}$. You cannot conclude, however, that the sample means are normally distributed unless either the underlying population is normally distributed or the sample size is large ($n \geq 30$).

Classroom Exercises

The following exercises provide more practice for investigating the sampling distribution for means of samples drawn from populations that are not normal.

1. a. Use **SetUpEditor** to restore the stat list editor to the standard lists L1 through L6. Clear L1, L2, and L3.

    b. In list L1, enter the following data.

    1, 2, 2, 3, 3, 3, 4, 4, 4, 4, 5, 5, 5, 5, 5, 6, 6, 6, 6, 6, 6, 7, 7, 7, 7, 7, 7, 7, 8, 8, 8, 8, 8, 8, 8, 8, 9, 9, 9, 9, 9, 9, 9, 9, 9, 0, 0, 0, 0, 0, 0, 0, 0, 0

    Notice that this fills L1 with one entry of 1, two entries of 2, three entries of 3, etc.

c. Verify that L1 contains 55 entries with mean approximately 5.1818 and standard deviation approximately 3.157.

2. Specify Plot1 to be a histogram with Xlist = L1 and Freq = 1. Let Xmin = −.5, Xmax = 9.5, and Xscl = 1. Does the population from which you will sample appear normal? Explain.

3. Modify the procedure listed on page 75 to obtain 100 sample means for samples of size four drawn from the population stored in L1. Store the sample means in L2. (Be sure to define C with an initial value of 1 and change the upper limit for the random integers to 55.) Calculate the mean and standard deviation of the list of sample means. How do these values compare to the values $\mu = 5.1818$ and $\sigma/\sqrt{4} = 1.5785$? Graph a histogram of the distribution. What can you conclude?

4. Modify the procedure listed on page 75 to obtain 100 sample means for samples of size twenty-five drawn from the population in L1. Store the sample means in L3. (Be sure to define C with an initial value of 1.) Calculate the mean and standard deviation of the list of sample means. How do these values compare to the values of $\mu = 5.1818$ and $\sigma/\sqrt{25} = 0.6314$? Graph a histogram of the distribution. How does the histogram differ from the one obtained in Exercise 3? What can you conclude?

5. Create normal probability plots for the sample means stored in L2 and L3. Does either list of sample means appear to be normally distributed? Explain.

14 Inference Procedures for Means (σ known)

This chapter provides several activities that allow students to perform and investigate the two most common types of statistical inference procedures: confidence intervals and significance tests. Specifically, the activities in this chapter describe how to use the TI-83 to

- create a confidence interval for a population mean.
- develop understanding of confidence intervals.
- perform a significance test for a population mean.
- investigate the effect of sample size on significance tests.

Each activity in this chapter uses means of samples drawn from a population having a known standard deviation. Before beginning the activities, you may want to review the following statements with your students.

- Sample means drawn from normal populations $N(\mu, \sigma)$ are normally distributed with mean equal to the population mean, μ, and standard deviation equal to $\sigma/\sqrt{n}$.
- Sample means drawn from any population are normally distributed with mean equal to the population mean, μ, and standard deviation equal to $\sigma/\sqrt{n}$, if the sample size is large ($n \geq 30$).

Creating a Confidence Interval for a Population Mean

This activity describes two methods for creating a confidence interval for a population mean. Consider the following problem.

> Assume that men's heights, in inches, are normally distributed with $\mu = 69$ and $\sigma = 2.5$, $N(69, 2.5)$. Suppose you sample twenty-five men and measure the height of each. Calculate the mean height of the men and use it create a 95% confidence interval for the population mean. Does your confidence interval contain the value 69, the population mean?

There are two ways to solve this problem using the TI-83. You can solve the problem using the definition of confidence interval for a mean or using the TI-83's statistical test features.

Using the Definition of Confidence Interval

To create a 95% confidence interval for the population mean of the men's heights using the definition of confidence interval, do the following.

1. Determine a sample mean height of the twenty-five men by executing the following command from the home screen. The sample mean is stored in M.

    mean(randNorm(69,2.5,25)) → M

2. Determine the margin of error associated with a 95% confidence interval. The standard deviation of these sample means is $2.5/\sqrt{25} = 0.5$. The z-value associated with a 95% confidence interval is determined by executing the command **invNorm(.975)**. The result is 1.96. Therefore, the margin of error associated with a 95% confidence interval is the product $(0.5)(1.96)$.

3. Determine the confidence interval by calculating its left and right endpoints. The left endpoint is $M - (0.5)(1.96)$. The right endpoint is $M + (0.5)(1.96)$.

Depending on the sample mean, M, the population mean may or may not be contained in the confidence interval.

Using the TI-83's Statistical Test Features

Another way to calculate a 95% confidence interval for the population mean of the men's heights is to use the TI-83's **ZInterval** feature. To calculate a 95% confidence interval using **ZInterval**, do the following.

1. Press [STAT] [◄] [7] to access the **ZInterval** menu.

2. Highlight Stats. Then enter the population standard deviation ($\sigma = 2.5$), the sample mean ($\bar{x} = M$), the sample size ($n = 25$), and the desired confidence level (C-level = .95). The calculator display should look similar to the one shown below.

> It is not necessary to calculate the sample mean before using **ZInterval** to determine the confidence interval. Instead, execute the command **randNorm(69,2.5,25) → L1** before accessing the **ZInterval** menu. Then highlight Data and define $\sigma = 2.5$, List = L1, Freq = 1, and C-level = .95. Highlight Calculate and press [ENTER]. The calculator will display the confidence interval, the sample mean, the sample standard deviation, and the sample size.

```
ZInterval
 Inpt:Data Stats
 σ:2.5
 x̄:68.978733610…
 n:25
 C-Level:.95
 Calculate
```

> If more decimal places are needed for the endpoints of the confidence interval, access the VARS statistics test menu by pressing [VARS] [5] [◄] [◄] then choose command **lower** ([ALPHA] [H]) or command **upper** ([ALPHA] [I]).

3. Highlight Calculate and press [ENTER]. The calculator will display the confidence interval, the sample mean, and the sample size.

Classroom Exercises

1. How wide is the confidence interval created in the activity? Suppose the sample size changes to 50. How does the width of the confidence interval change? How large a sample size is needed to cut the width of the original confidence interval in half?

2. Create a 90% confidence interval using the sample mean, M, obtained in the activity. How does the width of this interval compare to the width of the 95% confidence interval? Does the 90% confidence interval contain the population mean?

3. Create a 99% confidence interval using the sample mean, M, obtained in the activity. How does the width of this interval compare to the width of the 95% confidence interval? Does the 99% confidence interval contain the population mean?

Developing Understanding of Confidence Intervals

In this activity, students can use the TI-83's list features to develop understanding of the concept of a confidence interval. Consider the following problem.

> Raw SAT scores are scaled so that final scores are normally distributed with mean 500 and standard deviation 100. Generate one hundred random samples consisting of twenty-five SAT scores. Calculate the mean for each sample and use it to construct a 95% confidence interval for the population mean. Examine each confidence interval to determine how many contain the population mean. Record the value of the sample mean for each interval that does not contain the population mean. Then analyze the sample means that did not produce successful confidence intervals. Are these sample means surprisingly high or low, relative to what you would expect?

The most efficient way to solve this problem is to use the TI-83's list features as outlined below.

1. Use **SetUpEditor** to create lists named MEAN, LEFT, and RIGHT. List MEAN will contain the one hundred sample means, while lists LEFT and RIGHT will contain the left and right endpoints of the confidence intervals, respectively.

2. Execute the following command from the home screen to generate the one hundred sample means and store them in list MEAN. (See page 72 for an explanation of the command.)

    seq(mean(randNorm(500,100,25)),X,1,100,1) → LMEAN

3. Create the confidence intervals and store the left and right endpoints in lists LEFT and RIGHT, respectively. To calculate each endpoint using the sample means in list MEAN and store the results, attach the following formulas to lists LEFT and RIGHT.

    LEFT="LMEAN-1.96(100/5)"
    RIGHT="LMEAN+1.96(100/5)"

> *Remember:* A 95% confidence interval for μ is
> $\bar{x} \pm 1.96(\sigma/\sqrt{n})$.

4. Scroll through the lists. Count the number of intervals that do not contain 500, the population mean. Record the sample mean for each such interval. What percent of the intervals contain the population mean?

5. When analyzing the sample means for each confidence interval that does not contain the population mean, consider the distribution of sample means drawn from an underlying $N(500, 100)$ population with $n = 25$. These sample means are also normally distributed with mean μ and standard deviation $\sigma/\sqrt{n}$. That is, the sample means stored in list MEAN are a sample of size 100 from $N(500, 20)$. Use **1-Var Stats** to verify that the mean and standard deviation for the sample means in MEAN are consistent with this theory. You should notice that each sample mean whose confidence interval did not contain the population mean is approximately two or more standard deviations away from the population mean of 500. Calculate the z-scores associated with each unsuccessful sample mean. These z-scores indicate an unusually low or high sample mean. Therefore, it is not surprising that the confidence intervals they produce do not give accurate information about the population.

Classroom Exercises

1. Sort list MEAN in ascending order. (Because lists LEFT and RIGHT are defined using list MEAN, the elements in these lists are moved automatically with respect to the corresponding entry in list MEAN.) Scroll through the lists. Record the least sample mean and the greatest sample mean for which the associated confidence interval contains $\mu = 500$.

2. Do the sample means recorded in Exercise 1 produce successful 90% confidence intervals? Do they produce successful 99% confidence intervals? Investigate using **ZInterval**. Discuss your results.

3. Determine the width of the 95% confidence intervals.

4. Modify the formulas attached to lists LEFT and RIGHT to create 90% confidence intervals. How wide are the confidence intervals? Scroll through the lists and determine how many intervals contain $\mu = 500$.

5. Modify the formulas attached to lists LEFT and RIGHT to create 99% confidence intervals. How wide are the confidence intervals? Scroll through the lists and determine how many intervals contain $\mu = 500$.

6. Discuss the results of Exercises 1–5. How does the width of a confidence interval change as the confidence level increases?

Performing a Significance Test for a Population Mean

In this activity, students can use the TI-83's **DISTR** features to help them perform a significance test for a population mean. Then they can use the TI-83's **Z-Test** feature to perform the significance test automatically. Consider the following problem.

> A packing machine is set at 21 ounces. The weights of the packages filled by the machine are normally distributed with $\mu = 21$ ounces and $\sigma = 0.5$ ounces. Packages filled by the machine are routinely monitored by randomly selecting four packages and determining the mean weight of the package contents. One such test resulted in a mean weight of 20.4 ounces. Assuming the weighing apparatus is precise, is there sufficient evidence that the packing machine needs readjustment?

To solve this problem, you must determine whether a mean weight of 20.4 ounces is unusual. If the weights of the packages are normally distributed, $N(21, 0.5)$, the mean weight of four packages is $N(21, 0.25)$. What proportion of means from this distribution can you expect to be less than or equal to 20.4 ounces?

One way to determine this proportion is to use the TI-83's **ShadeNorm** command. Using **ShadeNorm** allows you to determine and graph the area under the normal curve associated with a specified mean value. For example, to determine the area under the normal curve $N(21, 0.25)$ associated with a sample mean less than or equal to 20.4, do the following.

1. Turn off or clear any existing stat plots, drawings, and Y= variables.

2. Define an appropriate viewing window. In this case, define Xmin = 19.5, Xmax = 22.5, Ymin = −.4, and Ymax = 1.7.

> Executing the command **normalcdf(0,20.4,21,.25)** calculates the same area as **ShadeNorm.** The disadvantage of using this command is the loss of the graphical representation.

3. Draw the normal curve and the associated area by executing the following command from the home screen. (See page 25 for more information on **ShadeNorm.**)

   ShadeNorm(0,20.4,21,.25)

   After executing the command, you should obtain a display similar to the one shown below. From the graph, you can determine that the area associated with a sample mean less than or equal to 20.4 is approximately 0.0082. Therefore, you can conclude that the sample mean is very low. In fact, if the machine is actually filling packages that are normally distributed with $\mu = 21$ and $\sigma = 0.5$, you can expect the mean weight of four packages to be less than or equal to 20.4 ounces less than 1% of the time. Machine readjustments should be considered.

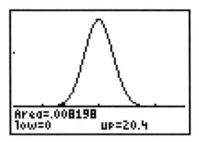

The method described above produces the *P*-value associated with a one-sided significance test. Another way to perform this test is to use the TI-83's **Z-Test** feature. To calculate the *P*-value associated with a one-sided significance test, do the following.

1. Press [STAT] [◄] [1] to access the **Z-Test** menu.

2. Highlight Stats. Then define $\mu_0 = 21$, $\sigma = .5$, $\bar{x} = 20.4$, $n = 4$, and $\mu < \mu_0$ for the alternative hypothesis. The calculator display should look similar to the one shown below.

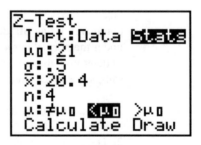

84   Chapter 14

3. To draw the graph and display the *P*-value, highlight Draw and press ENTER. The resulting display should be similar to the one shown below. Notice that the graph and the corresponding *z*- and *P*-values are displayed. The *P*-value is equal to the area determined using **ShadeNorm**.

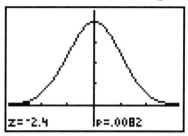

Instead of drawing the graph, you may simply want to calculate the *P*-value. To do so, highlight Calculate rather than Draw before pressing ENTER. The calculator will display the statement of the alternative hypothesis, the *z*- and *P*-values, the sample mean, and the sample size on the home screen.

Classroom Exercises

1. Suppose you had been asked to determine how likely it is to obtain a sample mean at least as extreme as 20.4 ounces when the packing machine is properly adjusted. Modify the procedure above to consider both exceptionally low and exceptionally high sample mean values. Modify the **Z-Test** procedure to perform a two-sided test. Discuss your results.

2. Using the sample mean 20.4 and population standard deviation 0.5, create a 95% confidence interval for the true value of μ associated with this machine. Does the confidence interval contain the hypothesized mean $\mu = 21$?

3. Would you consider an *individual* package unusual if it weighed 20.4 ounces? Explain.

Investigating the Effect of Sample Size on Significance Tests

In this activity, you will use the TI-83's **Z-Test** feature to investigate the effect of sample size on significance tests. Consider the following problem.

> Assume verbal SAT scores are normally distributed with mean 500 and standard deviation 100. An admissions officer at a certain university examines a random sample of applicants and calculates that the mean verbal SAT score for the sample is 530. Can the admissions officer conclude that the applicants to the university perform significantly better on the verbal test than the population of all college-bound students who take the test?

Before answering the question, you should consider the number of applicants in the sample. Without this information, the correct answer is, "It depends." For example, if the sample size is 100, you can use the **Z-Test** feature to determine a P-value of approximately 0.0013. Such a P-value indicates a highly significant result. Therefore, if the sample size is at least 100, the admissions officer can conclude that the applicants perform significantly better on the verbal SAT test than the population of students who take the test.

Suppose, however, that the sample size is only 25. Using the **Z-Test** feature, you can determine that the P-value increases to approximately 0.0668. Such a result is only moderately significant. Thus, the admissions officer may not want to make any conclusions concerning the applicants' verbal SAT test results.

Classroom Exercises

1. Suppose the admissions officer samples only ten applicants. Repeat the **Z-Test** using a sample size of 10. What is the P-value associated with this sample size? Interpret this P-value. What should the admissions officer conclude concerning the applicants' test scores? Explain.

2. Use the TI-83's **Z-Test** feature to investigate how different (larger or smaller) a sample mean must be from the population mean $\mu = 500$ to be significant at the 0.05 level when $n = 25$. What if $n = 250$? Use two-sided tests for these investigations.

15 Fixed Level Testing and Errors in Significance Tests

Most introductory statistics courses emphasize the role of *P*-values when conducting significance tests. Fixed level tests assess statistical significance at a prespecified level of significance, α. For example, if you test at the 0.05 level, you will reject H_0 if the *P*-value associated with the test is less than or equal to 0.05.

In this chapter, you will use the TI-83 to

- conduct fixed level tests.
- investigate Type I errors.
- investigate Type II errors.
- calculate the probability of a Type II error.

Conducting Fixed Level Tests

In this activity, you will use the TI-83 to conduct a fixed level test for a population mean when the standard deviation is known. Consider the following problem.

> A packing machine fills packages so that the content weights are normally distributed with mean 21 ounces and standard deviation 0.5 ounces. To monitor the machine, four packages are randomly selected and weighed. The mean content weight is then determined. If the sample mean differs significantly from 21 ounces at the 5% level ($\alpha = 0.05$), the machine is shut down and readjusted. For what sample means will the machine be shut down?

It is important to note that the goal in this problem is not to test a particular sample mean, but to determine all sample means which result in a machine shut down and a need for readjustment. To determine the tolerance limits, or acceptable weight bounds, for the machine, you must first determine the sampling distribution of the sample means. Because $n = 4$ and the underlying population is $N(\mu, \sigma)$, the sampling distribution of $\bar{x}$ is $N(\mu, \sigma/\sqrt{4})$, or $N(21, 0.25)$.

From this distribution, you can determine the values associated with the 2.5th percentile and with the 97.5th percentile. (These percentiles are chosen because they have a combined area of 0.05 to their left and right. Combining these areas is required because the two-sided alternative hypothesis suggests you reject H_0 when the sample mean differs significantly from 21 in either direction.) To calculate the values associated with each percentile, execute the following commands from the home screen.

    invNorm(.025,21,.25)   2.5th Percentile
    invNorm(.975,21,.25)   97.5th Percentile

You should obtain values of 20.51 and 21.49 for the 2.5th and 97.5th percentile, respectively. Therefore, the machine is shut down and readjusted whenever the mean contents of four randomly selected packages is less than or equal to 20.51 ounces or greater than or equal to 21.49 ounces.

The values 20.51 and 21.49 separate all possible sample means into two groups: one group suggests you reject the null hypothesis ($\bar{x} \leq 20.51$ or $\bar{x} \geq 21.49$) and the other suggests you do not reject the null hypothesis ($20.51 < \bar{x} < 21.49$).

Classroom Exercises

1. Use **ShadeNorm** to show graphically the acceptance and rejection regions determined above. (Define Xmin = 20, Xmax = 22, Ymin = −.2, and Ymax 1.7.) Verify that the area of the acceptance region is approximately 0.95.

2. Solve the problem using a significance level of 10% ($\alpha = 0.10$). For what sample means is the machine shut down and readjusted?

3. Solve the problem using a significance level of 1% ($\alpha = 0.01$). For what sample means is the machine shut down and readjusted?

4. How do the rejection values change as the significance level decreases? Increases? How do the acceptance values change in each case?

Investigating Type I Errors

Sometimes performing significance tests can lead to incorrect conclusions. For example, suppose you reject the null hypothesis when, in fact, it is true. Such an error is called a Type I error. For a test performed at a significance level α, the probability of making a Type I error is α.

To investigate the probability of Type I errors, consider the following problem.

> A packing machine fills packages so that the content weights are normally distributed with mean 21 ounces and standard deviation 0.5 ounces. Verify, using simulations, that the probability of making a Type I error is 0.05 when performing a significance test of $H_0: \mu_0 = 21$ versus $H_a: \mu \neq 21$ at the 5% level ($\alpha = 0.05$).

To simulate the sample mean of four packages from a $N(21, 0.5)$ population, execute

mean(randNorm(21,.5,4))

from the home screen. If the sample mean falls in the acceptance region ($20.51 < \bar{x} < 21.49$) the null hypothesis, $\mu_0 = 21$, is not rejected—a correct conclusion. However, if the sample mean falls in the rejection area ($\bar{x} \leq 20.51$ or $\bar{x} \geq 21.49$), the null hypothesis is rejected—an incorrect conclusion. Because each mean is determined using a random sample from the normal population $N(21, 0.5)$, any sample mean that results in rejecting the null hypothesis causes a Type I error.

One way to verify that the probability of making a Type I error is equal to the significance level $\alpha = 0.05$ is to calculate many sample means, count how many fall in the rejection area, and determine the percent of incorrect conclusions. You can use the TI-83 to accomplish these tasks as directed below.

1. Calculate one hundred sample means from the $N(21, 0.5)$ population and store the results in L1 by executing the following command from the home screen.

    seq(mean(randNorm(21,.5,4)),X,1,100,1)→L1

2. Use **SortA** to sort list L1 in ascending order. Count how many sample means are less than or equal to 20.51 or greater than or equal 21.49. Each such sample mean results in a Type I error.

3. Determine the percent of sample means that result in a Type I error.

The percent of sample means resulting in a Type I error should be approximately 0.05.

Classroom Exercises

1. For each sample mean in L1 less than or equal to 20.51 or greater than or equal to 21.49, use the TI-83's **Z-Test** feature to verify that the sample mean results in a *P*-value less than or equal to 0.05 when testing $H_0: \mu = 21$ versus $H_a: \mu \neq 21$.

2. Pool the results of others in your class. Then use the combined results to determine the percent of sample means that result in the incorrect rejection of H_0.

3. Use simulations to verify that the probability of a Type I error is 0.1 when a significance test is performed at the 10% level.

4. Use simulations to verify that the probability of a Type I error is 0.01 when a significance test is performed at the 1% level.

Investigating Type II Errors

When performing significance tests, a second type of error occurs if the decision is made not to reject the null hypothesis H_0 when, in fact, H_0 is not true. Such an error is called a Type II error. The Greek letter, β, represents the probability of committing a Type II error. To approximate the probability of a Type II error, β, consider the following problem.

> Suppose the packing machine's target setting changes from 21 ounces to 21.5 ounces, so that package weights are now distributed $N(21.5, 0.5)$. Use simulations to approximate the probability of making a Type II error, β, when testing $\mu = 21$ at the 5% level.

The goal of the problem is to approximate the probability that a sample mean of four packages selected from a $N(21.5, 0.5)$ population falls in the acceptance region $(20.51 < \bar{x} < 21.49)$ associated with $\mu = 21$. One way to approximate the probability of making a Type II error is to calculate many sample means, count how many fall in this acceptance area, and determine the percent of incorrect conclusions. You can use the TI-83 to accomplish these tasks as directed below.

1. Calculate one hundred sample means from the $N(21.5, 0.5)$ population and store the results in L2 by executing the following command from the home screen.

    seq(mean(randNorm(21.5,.5,4)),X,1,100,1) → L2

2. Use **SortA** to sort list L2 in ascending order. Count how many sample means fall into the acceptance region ($20.51 < \bar{x} < 21.49$) for $H_0: \mu = 21$ versus $H_a: \mu \neq 21$ when $\alpha = 0.05$. Each such sample mean results in a Type II error.

3. Determine the percent of sample means that result in a Type II error.

The percent of sample means resulting in a Type II error should be approximately 0.5. That is, $\beta \approx 0.5$.

Classroom Exercises

1. Determine the largest sample mean in L1 that results in an incorrect conclusion. Then use the TI-83's **Z-Test** feature to verify that the sample mean results in a P-value greater than 0.05 when testing $H_0: \mu = 21$ versus $H_a: \mu \neq 21$.

2. Pool the results of others in your class. Then use the combined results to determine the percent of sample means that suggest the incorrect acceptance of $\mu = 21$.

3. How does the probability of a Type II error change when the significance test is performed at the 10% level?

4. How does the probability of a Type II error change when the significance test is performed at the 1% level?

5. The probability of making a Type II error depends on the true value of the population mean. Perform simulations to verify that β decreases when the difference between the true value of μ and the hypothesized value of $\mu = 21$ increases.

Calculating the Probability of a Type II Error

Remember: The sample means are normally distributed with mean 21.5 and standard deviation $0.5/\sqrt{4} = 0.25$.

Using the TI-83's **normalcdf** command, you can calculate the probability of making a Type II error. To calculate β using the specifications in the packing-machine problem on page 90, that is, when the true value of μ is 21.5, execute

normalcdf(20.51,21.49,21.5,.25)

from the home screen. The result is 0.484. Therefore, if the true mean value is 21.5, you can expect to incorrectly accept the hypothesis that $\mu = 21$ about 48% percent of the time.

Classroom Exercises

1. Use **ShadeNorm** to represent β graphically. That is, determine the area between 20.51 and 21.49 for the normal distribution $N(21.5, 0.25)$. (Define Xmin = 20, Xmax = 23, Ymin = $-.4$, and Ymax = 1.7.)

2. Suppose the monitoring procedure for the packaging machine changes and ten packages are randomly selected instead of four. That is, the sample size increases to $n = 10$.

   a. How does the sampling distribution of sample means change? Explain.

   b. Let $\alpha = 0.05$. Determine the acceptance and rejection regions for a test of $\mu = 21$ versus $\mu \neq 21$.

   c. Determine the probability of making a Type II error when the true value of μ changes to 21.5.

16 The *t* Distribution

The *t* statistic is often used to make inferences about a population mean when sample sizes are small and the population standard deviation is unknown.

In this chapter, you will use the TI-83 to investigate *t* distributions. Specifically, you will use the TI-83 to

- graph density curves for *t* distributions.
- compare areas under the *t* and normal curves.
- calculate *t* values associated with percentiles.

Graphing Density Curves for t Distributions

In this activity, you will use the TI-83 to graph density curves for *t* distributions with various degrees of freedom. To help you better understand the graphs associated with the *t* distributions, you will compare each graph to the standard normal curve. Consider the following problem.

> Graph the standard normal curve and the density curves for *t* distributions with 2, 10, and 25 degrees of freedom. How does the density curve of the standard normal distribution compare with the density curves of *t*(2), *t*(10), and *t*(25) distributions?

To graph each density curve using the TI-83, do the following.

1. Clear any existing Y= variables, stat plots, and drawings.

2. Access the Y= editor. In Y1, enter the standard normal probability density function.

    Y1 = normalpdf(X)

3. Set an appropriate viewing window by defining Xmin = −4, Xmax = 4, and Xscl = 1. To define the remaining window settings, Ymin, Ymax, and Yscl, and graph the normal curve, use the TI-83's **ZoomFit** command by pressing [ZOOM] [0]. The calculator display should be similar to the one shown on the next page.

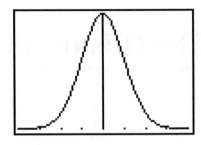

4. Access the Y= editor and define **Y2 = tpdf(X,2).** You can access command **tpdf** by pressing [2nd] [DISTR] [4]. When using **tpdf,** you must specify the degrees of freedom. Press [GRAPH] to display the standard normal curve and the $t(2)$ curve. The calculator display should be similar to the one shown below. Notice that both curves are bell-shaped and centered at 0, but the t distribution is spread out more than the standard normal distribution. This implies that the t distribution with 2 degrees of freedom has more area in the tails than the standard normal distribution.

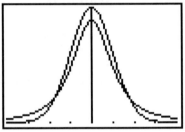

5. Access the Y= editor and define **Y3 = tpdf(X,10).** Press [GRAPH] to display the standard normal curve, the $t(2)$ curve, and the $t(10)$ curve. The calculator display should be similar to the one shown below. Notice that the density curve for the t distribution with 10 degrees of freedom is centered at 0 and lies between the standard normal curve and the density curve for the t distribution with 2 degrees of freedom.

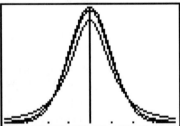

6. Access the Y= editor and turn off Y2 and Y3. Then define
   Y4 = tpdf(X,25). Press [GRAPH] to display the standard normal curve and the *t*(25) density curve. The calculator display should be similar to the one shown below. Using the present window settings, it is difficult to distinguish the two density curves. The density curve for the *t* distribution with 25 degrees of freedom is very similar to the standard normal curve.

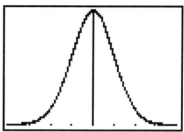

Comparing Areas Under the t and Normal Curves

In this activity, you will compare the area in the right tail of density curves for the standard normal distribution and for *t* distributions with various degrees of freedom. Using the TI-83, you can determine such areas graphically and numerically.

Comparing Areas Graphically

To compare areas associated with normal distributions and *t* distributions graphically, use the TI-83's **ShadeNorm** and **Shade_t** commands. To access each command, enter the following keystrokes.

[2nd] [DISTR] [▶] [1]   *Access **ShadeNorm** command.*
[2nd] [DISTR] [▶] [2]   *Access **Shade_t** command.*

Consider the following problem.

> Determine the area to the right of $z = 1$ under the standard normal curve. Then determine the area to the right of $t = 1$ under the density curves for *t* distributions with 2, 10, and 25 degrees of freedom. What can you conclude?

To determine each desired area graphically using the TI-83, do the following.

1. Turn off any existing Y= variables, stat plots, and drawings.

2. Define Xmin = 0, Xmax = 4, Xscl = 1, Ymin = −.1, and Ymax = .4. Defining Ymin = −.1 assures the graph captions do not interfere with the graphs.

3. Determine the area to the right of $z = 1$ under the standard normal curve by executing **ShadeNorm(1,10^9)** from the home screen. The values 1 and 10^9 specify the lower and upper bounds, respectively. The calculator display should be similar to the one shown below. Notice that the area to the right of $z = 1$ for the standard normal distribution is approximately 0.158655.

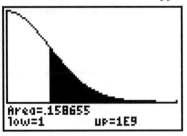

4. Use **ClrDraw** to clear the existing drawing. Then determine the area to the right of $t = 1$ under the density curve for the t distribution with 2 degrees of freedom by executing **Shade_t(1,10^9,2)** from the home screen. The values 1, 10^9, and 2 specify the lower bound, the upper bound, and the degrees of freedom, respectively. The resulting display should look similar to the one shown below. Notice that the area to the right of $t = 1$ for the $t(2)$ distribution is approximately 0.211325.

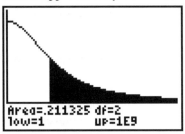

5. In a similar manner, determine the area to the right of $t = 1$ under the density curve for the t distributions with 10 and 25 degrees of freedom by executing **Shade_t(1,10^9,10)** and **Shade_t(1,10^9,25)** from the home screen. The resulting calculator displays for these commands should look similar to the ones shown below. Notice that the area to the right of $t = 1$ for the $t(10)$ distribution is approximately 0.170447 and that the area to the right of $t = 1$ for the $t(25)$ distribution is approximately 0.163446.

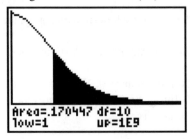 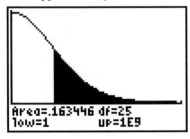

You should notice from the graphs that as the degrees of freedom increase, there is less area under the graph to the right of $t = 1$. You should also notice that as the degrees of freedom increase, the areas associated with each t distribution approach the corresponding area associated with the standard normal distribution. In fact, the area to the right of $t = 1$ under the density curve for the t distribution with 25 degrees of freedom differs from the area to the right of $z = 1$ under the standard normal curve by less than 0.005.

Comparing Areas Numerically

To compare areas associated with normal distributions and t distributions numerically, use the TI-83's **normalcdf** and **tcdf** commands. To access each command, enter the following keystrokes.

[2nd] [DISTR] [2]   *Access **normalcdf** command.*
[2nd] [DISTR] [5]   *Access **tcdf** command.*

Consider the following problem.

> Determine the area to the right of $z = 1$ under the standard normal curve. Then determine the area to the right of $t = 1$ under the density curve for the t distribution with 100 degrees of freedom. What can you conclude?

To determine each desired area using the TI-83, execute the following commands from the home screen. Note that you must specify lower and upper bounds for each command. You must also specify the degrees of freedom for the t distribution.

normalcdf(1,10^9)
tcdf(1,10^9,100)

After entering the commands, observe that the area to the right of $z = 1$ for the standard normal distribution is approximately 0.158655 while the area to the right of $t = 1$ for the $t(100)$ distribution is approximately 0.159862. These areas differ by only 0.0012.

Classroom Exercises

1. Use **ShadeNorm** and **Shade_t** to compare areas to the right of $z = 2$ and $t = 2$ under the standard normal curve and the density curves for t distributions with 2, 10, and 25 degrees of freedom.

2. Use **normalcdf** and **tcdf** to compare areas to the right of $z = 3$ and $t = 3$ under the standard normal curve and the density curves for t distributions with 2, 10, and 25 degrees of freedom.

3. For each problem discussed in this activity, an upper bound of 10^9 is used. If you specify a larger upper bound, how will the calculated area change? Investigate by increasing the upper bound. What can you conclude? (The TI-83 Guidebook suggests using 1E99 to represent infinity.)

4. Execute each command listed below from the home screen. What area is represented by the result of each command?

   a. $0.5 - \text{normalcdf}(0, 1)$

   b. $0.5 - \text{tcdf}(0, 1, 2)$

5. Explain how to use the results of Exercise 4 to find the exact area to the right of $t = 1$ for the t distribution with 100 degrees of freedom.

Calculating t Values Associated with Percentiles

Because the TI-83 has no inverse t feature that parallels the **invNorm** command, you must use the calculator's equation solver to find a t distribution percentile. Consider the following problem.

Find the 95th percentile for the *t* distribution with 2 degrees of freedom.

To find the 95th percentile for this distribution, you must solve the equation

$$\text{tcdf}(-1\text{E}99, T, 2) - .95 = 0.$$

In the equation, the values $-1\text{E}99$, 2, and .95 represent the lower bound ($-\infty$), the degrees of freedom, and the percentile, respectively. To solve the equation using the TI-83's equation solver, enter the following keystrokes.

Keystrokes	Description
[MATH] [0]	*Display equation solver.*
[▲] [CLEAR]	*Clear equation solver, if necessary.*
[2nd] [DISTR] [5]	*Enter equation.*
[(-)] 1 [2nd] [EE] 99 [,]	
[ALPHA] [T] [,] 2 [)]	
[−] .95	
[ENTER]	
3	*Enter initial guess.*
[ALPHA] [SOLVE]	*Solve equation.*

Be sure the cursor is on the T = entry line before pressing [ALPHA] [SOLVE].

The resulting value of t is approximately 2.919986. Therefore, you can conclude that about 95% of the area under the density curve for the t distribution with 2 degrees of freedom is to the left of $t = 2.92$.

Classroom Exercises

1. Find the 5th and 99th percentiles for the t distribution with 2 degrees of freedom.

2. Find the 2.5th and 97.5th percentiles for the t distribution with 10 degrees of freedom.

3. Find the interquartile range for the t distribution with 25 degrees of freedom.

4. Use the equation solver to solve the equation $0 = \text{tcdf}(0,T,2) - .45$. What does the solution to this equation represent?

17 One-Sample and Matched Pairs *t* Procedures

In Chapters 14 and 15, several methods were discussed for creating confidence intervals and performing significance tests for a population mean, μ, when the population standard deviation, σ, is known. More often than not, however, the population standard deviation is *not* known and the sample standard deviation, s, is used to estimate σ. If this is the case, the statistic

$$\frac{\bar{x} - \mu}{s/\sqrt{n}}$$

has the t distribution with $n - 1$ degrees of freedom provided $\bar{x}$ and s are based on a simple random sample of size n selected from a normally distributed population.

In this chapter, you will use the TI-83 to

- create a one-sample t confidence interval.
- perform a one-sample t significance test.
- analyze matched pairs.

Creating a One-Sample *t* Confidence Interval

This activity describes two methods for using the TI-83 to create a one-sample t confidence interval. For example, you can create such a confidence interval using the definition of confidence interval. Or you can use the TI-83's statistical test features. Consider the following problem.

> In a study of cholesterol levels, blood samples are taken from an individual for five consecutive days. The samples are then analyzed and the cholesterol levels reported. The reported daily cholesterol levels for one individual are 228, 235, 238, 226, and 230. Use these data to determine a 95% confidence interval for the individual's cholesterol level.

Using the Definition of Confidence Interval

Assuming the data are a simple random sample from a normally distributed population, a one-sample t confidence interval for the population mean, μ, is given by

$$\bar{x} \pm t^* \frac{s}{\sqrt{n}}$$

where $\bar{x}$ is the sample mean, s is the sample standard deviation, n is the sample size, and t^* represents the appropriate t-value associated with the $t(n-1)$ distribution and specified confidence level.

To create a 95% confidence interval for the individual's true cholesterol level, do the following.

1. Use **SetUpEditor** to restore the standard lists, L1 through L6, in the stat list editor. Then clear list L1 and enter the cholesterol data into L1.

2. Calculate the sample mean and the sample standard deviation for the data in L1. Using **1-Var Stats,** you should obtain $\bar{x} = 231.4$ and $s \approx 4.98$.

3. Determine t^*. In this problem, t^* is the 97.5th percentile of the t distribution with 4 degrees of freedom. To determine t^*, use the TI-83's equation solver to solve the equation $0 = \text{tcdf}(-1E99, T, 4) - .975$. (For more information on using the TI-83's equation solver, see page 98.) Specify an initial guess of $t = 2$. Using the equation solver, you can determine that $t^* = T \approx 2.7764$.

4. Determine the confidence interval by calculating the left and right endpoints. The left endpoint is $\bar{x} - T\frac{s}{\sqrt{n}}$. The right endpoint is $\bar{x} + T\frac{s}{\sqrt{n}}$. You can approximate the endpoints using rounded values for T and s or you can find the exact endpoints by executing the following commands from the home screen.

mean(L1)-T*stdDev(L1)/√(5)	Left endpoint
mean(L1)+T*stdDev(L1)/√(5)	Right endpoint

   You should conclude that the 95% confidence interval is (225.22, 237.58).

Using the TI-83's Statistical Test Features

Another way to create a 95% confidence interval for the individual's true cholesterol level is to use the TI-83's **TInterval** feature. To create a 95% confidence interval using **TInterval**, do the following.

1. Press [STAT] [◄] [8] to access the **TInterval** menu.

2. Highlight Data. Then define List = L1, Freq = 1, and C-Level = .95. The calculator display should look similar to the one shown below.

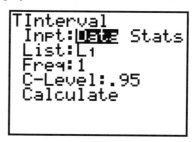

3. Highlight Calculate and press [ENTER]. The calculator will display the confidence interval, the sample mean, the sample standard deviation, and the sample size.

Classroom Exercises

1. As the sample size changes, how does the width of the 95% confidence interval change? Investigate by highlighting Stats under the **TInterval** menu and changing the value of n. Do not change the sample mean or the sample standard deviation. Discuss your results.

2. Create a 99% confidence interval for the individual's true cholesterol level. Use each method discussed in the activity. How does the width of this interval compare to the width of the 95% confidence interval?

3. Create a 90% confidence interval for the individual's true cholesterol level. Use each method discussed in the activity. How does the width of this interval compare to the width of the 95% and 99% confidence intervals?

Performing a One-Sample t Significance Test

In this activity, you can use TI-83's **tcdf** command to help your students perform a one-sample t significance test. Then using the TI-83's **T-Test** feature, you can show your students how to perform a significance test automatically. Consider the following problem.

> In a study of cholesterol levels, blood samples are taken from an individual for five consecutive days. The samples are then analyzed and the cholesterol levels reported. The reported daily cholesterol levels for one individual are 228, 235, 238, 226, and 230. Researchers specify that subjects in the study should have a cholesterol level of 230. Should this individual be allowed to participate?

To solve this problem, you can base your decision on a significance test with the hypotheses $H_o: \mu = 230$ and $H_a: \mu \neq 230$. Using the data in list L1 and the **1-Var Stats** command, you can determine that the sample mean is $\bar{x} = 231.4$ and the sample standard deviation is $s \approx 4.98$. Therefore, the test statistic, t, with $n - 1 = 4$ degrees of freedom is

$$t = \frac{\bar{x} - \mu}{s/\sqrt{n}} \approx \frac{231.4 - 230}{4.98/\sqrt{5}} \approx 0.63.$$

> Execute the following command to obtain an exact value for t.
> **(mean(L1)-230) / (stdDev(L1)/√(5)) → T**
> Then execute the command
> **1 − tcdf(−T,T,4)**
> to obtain an exact P-value.

Because this is a two-sided test, the P-value associated with $t = 0.63$ includes the area to the left of -0.63 and the area to the right of 0.63. To calculate this area, execute the command

1 − tcdf(−.63, .63, 4)

from the home screen. The resulting P-value is approximately 0.5629. From this value, you can conclude that the individual's cholesterol levels are reasonable for a person with a cholesterol level of 230 and the individual should be allowed to participate in the study.

Another way to perform the two-sided significance test with $H_o: \mu = 230$ and $H_a: \mu \neq 230$ is to use the TI-83's **T-Test** feature. To calculate the P-value automatically, do the following.

1. Press [STAT] [◄] [2] to access the **T-Test** menu.

2. Highlight Data. Then define $\mu_o = 230$, List = L1, and Freq = 1. Because this is a two-sided test, define $\mu \neq \mu_o$ for the alternative hypothesis.

3. To draw the graph of the t distribution with 4 degrees of freedom and display the area associated with the P-value, highlight Draw and press [ENTER]. The resulting display should be similar to the one shown on the next page. Notice the corresponding t- and P-values are displayed.

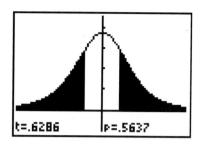

Instead of drawing the graph, you can simply calculate the *P*-value by highlighting Calculate rather than Draw before pressing ENTER. The calculator will display the statement of the alternative hypothesis, the *t*- and *P*-values, the sample mean, the sample standard deviation, and the sample size.

Classroom Exercises

1. As the sample size changes, how do the test statistic, *t*, and the *P*-value associated with the two-sided significance test change? Investigate by highlighting Stats under the **T-Test** menu and changing the value of *n*. Do not change the sample mean or the sample standard deviation. Discuss your results.

2. Assuming the sample mean and the sample standard deviation do not change, how large a sample size is needed to reject the null hypothesis at the 0.05 level? Explain your reasoning.

Analyzing Matched Pairs

In this activity, you will use the TI-83's list and statistical test features to analyze matched pairs. Consider the following problem.

> The final exam for a certain course is always placed last in the exam schedule. The instructor of this course questions whether this placement has an adverse affect on the final exam grades. The instructor hypothesizes that students become "burned out" by the final day of exams and their grades tend to fall. To test this hypothesis, the instructor wants to compare the midterm exam grade (which is placed early in the exam schedule) with the final exam grade for fifteen randomly selected students in the course. The midterm and final exam grades for these students are listed on the next page.

Student	Midterm Exam	Final Exam
1	97	84
2	98	92
3	90	83
4	71	67
5	87	78
6	91	81
7	95	96
8	75	78
9	87	83
10	95	89
11	84	82
12	69	61
13	79	80
14	95	94
15	92	88

Table 17.1

To analyze the change, or difference, in the matched scores, do the following.

1. Use **SetUpEditor** to restore the standard lists L1 through L6 in the stat list editor. Clear lists L1, L2, and L3.

2. Enter the midterm and final exam grades in L1 and L2, respectively. Then define L3 = L2 − L1. Note that the entries in list L3 indicate the difference in the exam scores. Scrolling through L3, you can see that most of the entries are negative, indicating a decrease in exam grades.

3. Create a box plot or a histogram using the data in L3 to check that the data are not strongly skewed and do not contain outliers. Because the distribution of differences is reasonably symmetric without any apparent skewness or outliers, it is acceptable to use a t test.

4. Perform a one-sided significance test using the hypotheses $H_o: \mu = 0$ and $H_a: \mu < 0$ to assess whether the final exam grades are lower than the midterm exam grades. The null hypothesis states that there is no change in grades while the alternative hypothesis states that the final exam grades are lower than the midterm exam grades. To perform the significance test, access the **T-Test** menu, highlight Data, and define $\mu_o = 0$, List = L3, Freq = 1, and $\mu < \mu_o$ for the alternative hypothesis. Then highlight Calculate and press [ENTER]. The results are displayed on the home screen. For this test, $t \approx -3.97$, $P \approx 0.00069$, $\bar{x} = -4.6$, $s \approx 4.48$.

The very small *P*-value indicates that the decrease in exam grades is very unlikely to result from chance alone. There is no guarantee, however, that the decrease is a result of the exam schedule. For example, it may be that the final exam is more difficult than the midterm exam. Whatever the reason, the differences are great enough to conclude that the final exam grades did decrease more than expected simply by chance.

Classroom Exercises

1. Create a normal probability plot of the data in L3. Does the plot indicate a distribution that is approximately normal? Explain.

2. Create a 95% confidence interval for the mean difference in scores of all the students who took both exams. Does this interval contain 0?

18 Two-Sample *t* Procedures

In this chapter, you will use the TI-83 to investigate the difference between two population means. Specifically, you will use the TI-83 to

- perform a two-sample *t* significance test.
- create a two-sample *t* confidence interval.

To perform these tasks, the data being analyzed should be independent simple random samples selected from normal populations. If so, statistical theory states that the sampling distribution of the difference in sample means, $\bar{x}_1 - \bar{x}_2$, is also normal. Because the population standard deviations, σ_1 and σ_2, are usually unknown, you can estimate each using the sample standard deviations, s_1 and s_2. Inference procedures are then based on the two-sample *t* statistic

$$t = \frac{(\bar{x}_1 - \bar{x}_2) - (\mu_1 - \mu_2)}{\sqrt{\frac{s_1^2}{n_1} + \frac{s_2^2}{n_2}}}.$$

Often, students in an introductory statistics course do not have easy access to technology with inferential statistics functions. Thus, most textbooks describe how to use critical values from a *t* distribution with degrees of freedom equal to the smaller of $n_1 - 1$ and $n_2 - 1$. Any results obtained using these degrees of freedom are conservative. For example, the confidence intervals created using these degrees of freedom have higher confidence than specified and any *P*-values that result from significance tests using these degrees of freedom are higher than the true *P*-values. Using the TI-83 to perform two-sample *t* procedures, however, allows students to use more accurate degrees of freedom calculated from the data. Refer to the TI-83 Guidebook (page A-53) or to a statistics textbook for the degrees of freedom formula.

The two-sample *t* statistic given above does not have an exact *t* distribution. Rather, the distribution is approximated by a *t* distribution and the approximation is more accurate when the degrees of freedom are determined using the data. If you can assume that the unknown population standard deviations are equal, then the pooled two-sample *t* statistic can be used for inferential procedures. The pooled two-sample *t* statistic has an exact *t* distribution with $n_1 + n_2 - 2$ degrees of freedom. The TI-83 prompts users to decide whether to pool the variances. Because of the TI-83's method for calculating degrees of freedom and the difficulty of verifying equal variances, the examples in this chapter do not discuss the pooled option. It is investigated, however, in the classroom exercises.

Performing a Two-Sample t Significance Test

In this activity, you can use the TI-83's **tcdf** command to help your students perform a two-sample t significance test. Then using the TI-83's **2-SampTTest** feature, you can show your students how to perform a two-sample t significance test automatically. Consider the following problem.

> You make two paper airplanes using a different model for each plane and want to determine whether the planes fly equal distances. You launch each plane ten times, alternating between planes, and measure the distance of each flight. (Assume ideal conditions for the flights—no arm fatigue, constant point of release, no wear and tear on the planes, etc.) The flight distances, in inches, for each plane are listed below.
>
> Plane A: 109, 150, 144, 169, 133, 145, 177, 160, 127, 166
> Plane B: 135, 157, 139, 152, 104, 97, 140, 129, 116, 117
>
> Based on these data, is there significant evidence at the 0.10 level that one plane flies farther than the other?

One way to help your students understand the concept of a two-sample t significance test is to demonstrate the "traditional" method for performing such tests and simply use the TI-83 to perform any necessary calculations. For example, to test whether the mean distance flown by Plane A is different from the mean distance flown by Plane B, do the following.

1. Use **SetUpEditor** to create two lists named PLNA and PLNB. Then enter the flight data into the appropriate list.

2. Create normal probability plots to verify that the data for each plane are approximately normal.

3. Perform a two-sample t test with hypotheses $H_o: \mu_1 = \mu_2$ and $H_a: \mu_1 \neq \mu_2$. You must perform a two-sided test to determine whether the planes flew different distances. To obtain the summary statistics for each plane's flight data, execute the command **2-Var Stats LPLNA, LPLNB**. You should obtain $\bar{x}_1 = 148$, $\bar{x}_2 = 128.6$, $s_1 \approx 20.992$, and $s_2 \approx 19.806$. Therefore, the t statistic is

$$t \approx \frac{(148 - 128.6) - (0)}{\sqrt{\frac{20.992^2}{10} + \frac{19.806^2}{10}}} \approx 2.126.$$

Because this is a two-sided test, the conservative P-value associated with $t = 2.126$ includes the area to the left of -2.126 and the area to the right of 2.126 under the $t(9)$ density curve. To find this P-value, execute the command

1 − tcdf(−2.126,2.126,9)

from the home screen. The result is approximately 0.0624. Because this P-value is less than the specified level of 0.10, you can conclude that there is a difference in the mean flight distances of the planes.

Another way to perform a two-sample t significance test is to use the TI-83's **2-SampTTest** feature. For example, to perform a significance test similar to one performed above but using more accurate degrees of freedom, do the following.

1. Press [STAT] [◄] [4] to access the **2-SampTTest** menu.

2. Highlight Data. Then define List1 = PLNA, List2 = PLNB, Freq1 = 1, and Freq2 = 1. Because this is a two-sided test, define $\mu1 \neq \mu2$ for the alternative hypothesis. Highlight No next to Pooled to specify unpooled variances.

3. Highlight Calculate and press [ENTER] to display the test results $t \approx 2.126$ and $P \approx 0.048$, the degrees of freedom $df \approx 17.94$, and the sample mean, sample standard deviation, and sample size for each set of data. Notice that the P-value and degrees of freedom are different from the previous test. Because of the greater degrees of freedom, the resulting P-value even more strongly supports the conclusion that there is a difference in the mean flight distances of the planes.

To provide a graphical representation of the significance test, highlight Draw rather than Calculate before pressing [ENTER]. The resulting display should be similar to the one shown below. Notice that only the t and P values are displayed.

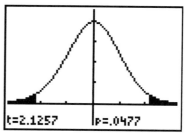

Classroom Exercises

1. Verify the P-value, 0.0477, obtained using the **2-SampTTest** feature by executing the following command from the home screen.

    $1 - \text{tcdf}(-2.1257, 2.1257, 17.9)$

2. Create parallel box plots for the flight distances of Plane A and Plane B. Use the box plots to explain why it might be reasonable to perform a

one-sided test to determine whether the mean flight distance of Plane A is significantly greater than the mean flight distance of Plane B. Based on the results of the activity, what *P*-value is associated with this one-sided alternative? Use the **2-SampTTest** feature to confirm your answer.

3. Because the sample standard deviations for Plane A and Plane B are very close in value, it may be reasonable to assume that their population variances are equal and to use the pooled two-sample *t* statistic when performing the significance test. Repeat the significance test using the pooled option. How do the resulting degrees of freedom and *P*-value compare with those obtained in the activity?

Creating a Two-Sample t Confidence Interval

This activity describes two ways to use the TI-83 to create a two-sample *t* confidence interval. For example, you can create such confidence intervals using the definition of a two-sample confidence interval and simply use the TI-83 to perform tedious calculations, or you can use the TI-83 to create the confidence interval automatically. Consider the following problem.

> You make two paper airplanes using a different model for each plane and launch each plane ten times, alternating between planes, and measure the distance of each flight. (Assume ideal conditions for the flights—no arm fatigue, constant point of release, no wear and tear on the planes, etc.) The flight distances, in inches, for each plane are listed below.
>
> Plane A: 109, 150, 144, 169, 133, 145, 177, 160, 127, 166
>
> Plane B: 135, 157, 139, 152, 104, 97, 140, 129, 116, 117
>
> Create a 90% confidence interval for the difference in mean flight distances for Plane A and Plane B.

One way to create this 90% confidence interval is to use the definition of a two-sample *t* confidence interval. Assuming the data sets consist of independent simple random samples selected from normal populations, a two-sample *t* confidence interval for $\mu_1 - \mu_2$ is given by

$$(\bar{x}_1 - \bar{x}_2) \pm t^* \sqrt{\frac{s_1^2}{n_1} + \frac{s_2^2}{n_2}}$$

where $\bar{x}_1$, s_1, and n_1 are the sample mean, sample standard deviation, and sample size for one data set, and $\bar{x}_2$, s_2, and n_2 are the sample mean, sample standard deviation, and sample size for the second data set. The t^* value represents the appropriate *t*-value associated with the specified confidence level and the *t* distribution with $n_1 - 1$ or $n_2 - 1$ degrees of freedom, whichever is less (a conservative approach).

To create the 90% confidence interval for the difference in mean distances flown by the planes, make sure the data are entered in lists PLNA and PLNB and do the following.

1. Calculate the sample statistics for list PLNA and PLNB. You should obtain $\bar{x}_1 = 148$, $\bar{x}_2 = 128.6$, $s_1 \approx 20.992$, $s_2 \approx 19.806$.

2. Determine t^*. In this problem, t^* is the 95th percentile of the t distribution with 9 degrees of freedom. To calculate t^*, use the TI-83's equation solver to solve the equation $0 = \text{tcdf}(-1\text{E}99, T, 9) - .95$. (For more information on the TI-83's equation solver, see page 98.) Specify an initial guess of 2. Using the equation solver, you should determine that $t^* = T \approx 1.833$.

3. Determine the confidence interval by calculating the left and right endpoints. The left and right endpoints are

$$(\bar{x}_1 - \bar{x}_2) \pm T \sqrt{\frac{s_1^2}{n_1} + \frac{s_2^2}{n_2}}$$

$$\approx (148 - 128.6) \pm 1.833 \sqrt{\frac{20.992^2}{10} + \frac{19.806^2}{10}}$$

$$\approx 19.4 \pm 16.73$$

Therefore, the 90% confidence interval is (2.67, 36.13).

You can create a two-sample t confidence interval automatically using the TI-83. For example, to create the 90% confidence interval for the difference in mean flight distances, do the following.

1. Press [STAT] [◄] [0] to access the **2-SampTInt** menu.

2. Highlight Data. Then define List1 = PLNA, List2 = PLNB, Freq1 = 1, Freq2 = 1, and C-Level = .90. Highlight No next to Pooled to specify unpooled variances.

3. Highlight Calculate and press [ENTER]. The calculator will display the confidence interval (3.5711, 35.229), the degrees of freedom $df \approx 17.939$, and various summary statistics. Note that because of the more accurate degrees of freedom, the confidence interval is not as wide as the confidence interval created using 9 degrees of freedom.

Classroom Exercises

1. Explain, using the results of the first activity, why it is expected that the confidence interval does not contain 0.

2. Use the equation solver to find the t^* value associated with a 90% confidence interval and 17.9 degrees of freedom. Then use the result to verify the confidence interval calculated using the **2-SampTInt** feature.

3. Do you expect a 99% confidence interval for the difference in mean flight distances of Plane A and Plane B to contain 0? Explain. Then confirm your answer using the methods discussed in the activity.

4. Calculate the 90% confidence interval using the pooled option. How do the pooled results compare to the unpooled results?

19 Inference for Population Proportions

This chapter provides several activities that enable students to investigate inference procedures for population proportions. Specifically, the activities in this chapter describe how to use the TI-83 to

- create a confidence interval for a population proportion.
- develop understanding of confidence intervals for population proportions.
- perform a significance test for a population proportion.

Before beginning the activities, make sure your students understand the sampling distribution of a sample proportion, $\hat{p}$. Specifically, for large sample sizes, the distribution of $\hat{p}$ is approximately normal with mean equal to the population proportion p and standard deviation equal to $\sqrt{p(1-p)/n}$.

Creating a Confidence Interval for a Population Proportion

This activity describes two ways to use the TI-83 to create a confidence interval for a population proportion. You can create such a confidence interval using the definition of confidence interval and simply use the TI-83 to perform tedious calculations, or you can use the TI-83 to create a confidence interval automatically. Consider the following problem.

> Use the TI-83 to simulate taking a simple random sample of 100 voters from a precinct in which 60% of the voting population favor a certain candidate. Calculate the sample proportion of voters who favor the candidate and use this value to create a 95% confidence interval for the population proportion. Does your confidence interval contain the population proportion?

To simulate taking a simple random sample of 100 voters from a population in which 60% favor a certain candidate and calculate the sample proportion of voters who favor that candidate, execute the following command from the home screen.

randBin(100,.6)/100

The resulting value is $\hat{p}$. For this activity, assume $\hat{p} = 0.58$.

One way to create a 95% confidence interval for the population proportion p is to use the definition of confidence interval. By definition, the confidence interval for p is

$$\hat{p} \pm z^* \sqrt{\frac{\hat{p}(1-\hat{p})}{n}}$$

where $\hat{p}$ is the sample proportion, n is the sample size, and z^* represents the standard normal value with an area of $(1 - 0.95)/2 = 0.025$ to its right.

To create the 95% confidence interval for the population proportion of voters who favor the candidate, do the following.

1. Determine $\hat{p}$ using the command **randBin(100,.6)/100**. (Assume $\hat{p} = 0.58$.)

2. Determine z^*. The z^*-value corresponding to a 95% confidence level is approximately 1.96. You can calculate this value using the command **invNorm(.975)**. (See page 29 for more information on **invNorm**.)

3. Determine the confidence interval by calculating the left and right endpoints. The left and right endpoints are

$$\hat{p} \pm z^* \sqrt{\frac{\hat{p}(1-\hat{p})}{n}} \approx 0.58 \pm 1.96 \sqrt{\frac{0.58(1-0.58)}{100}}$$

$$\approx 0.58 \pm 0.097.$$

Therefore, the confidence interval is (0.483, 0.677). This interval contains the true population proportion $p = 0.60$.

You can create a confidence interval for a population proportion automatically using the TI-83. For example, to create a 95% confidence interval for the population proportion of voters who favor the candidate, do the following.

1. Press STAT ◄ ALPHA [A] to access the **1-PropZInt** menu.

2. The **1-PropZInt** menu prompts you to enter three values: x, n, and C-Level. The x value represents the number of successes, n represents the sample size, and C-Level represents the level of confidence. For this example, define $x = 58$, $n = 100$, and C-Level $= .95$.

3. Highlight Calculate and press ENTER. The calculator will display the confidence interval (.48326, .67674), $\hat{p} = 0.58$, and $n = 100$.

Classroom Exercises

1. Determine the width of the 95% confidence interval created in the activity. How does the width change if the sample size decreases to 50 and the sample proportion remains unchanged? How does the width change if the sample size increases to 200 and the sample proportion remains unchanged? How large a sample size is needed to cut the width in half?

2. Create a 90% confidence interval for the population proportion p using $\hat{p} = 0.58$. How does the width of this confidence interval compare to the width of the 95% confidence interval? Does the confidence interval contain the population proportion p?

3. Create a 99% confidence interval for the population proportion using $\hat{p} = 0.58$. How does the width of this confidence interval compare to the width of the 95% and 90% confidence intervals? Does the confidence interval contain the population proportion p?

4. The TI-83's **Line** command enables you to display a confidence interval graphically. For example, to display the 95% confidence interval created in the activity, do the following.

    a. Turn off or clear any Y= variables, stat plots, and drawings.

    b. Define Xmin = .4, Xmax = .8, Xscl = .1, Ymin = 0, Ymax = 20, and Yscl = 5.

    c. Enter the following keystrokes from the home screen.

    | 2nd [DRAW] 2 | *Access* **Line** *command.* |
    | .483 , 10 , | *Define coordinates of left endpoint.* |
    | .677 , 10) | *Define coordinates of right endpoint.* |
    | ENTER | |

    Use *y*-coordinates of 5 and 15 to display graphically the 95% confidence intervals created in Exercise 1 for samples of size 50 and 200. Discuss your results.

5. Use the TI-83's **Line** command to display graphically the 90%, 95%, and 99% confidence intervals based on samples of size 100. Discuss your results.

Developing Understanding of Confidence Intervals for Population Proportions

In this activity, students can use the TI-83's list features to help them understand the concept of a confidence interval for a population proportion. Consider the following problem.

> Use the TI-83 to simulate taking one hundred random samples consisting of 100 voters from a precinct where 60% of the voters favor a certain candidate. For each sample, calculate $\hat{p}$ and create a 95% confidence interval for the population proportion. Examine each confidence interval to determine how many contain the population proportion $p = 0.60$. Record the value of the sample proportions for those intervals that do not contain the population proportion. Then analyze the sample proportions that did not produce a successful confidence interval. Are these sample proportions surprisingly high or low, relative to what you would expect? Explain.

The most efficient way to solve this problem is to use the TI-83's list features as described below.

1. Use **SetUpEditor** to create lists named PHAT, PLOW, and PHIGH. List PHAT will contain the one hundred sample proportions, while lists PLOW and PHIGH will contain the left and right endpoints of the confidence intervals, respectively.

2. Execute the following command from the home screen to generate the one hundred sample proportions and store them in list PHAT.

    randBin(100,.6,100)/100 → LPHAT

3. Create the confidence intervals and store the left and right endpoints in lists PLOW and PHIGH, respectively. To calculate each endpoint using the sample proportions in list PHAT and store the results, attach the following formulas to lists PLOW and PHIGH.

    PLOW = "LPHAT − 1.96 × $\sqrt{\text{LPHAT} \times (1 - \text{LPHAT}/100)}$"
    PHIGH = "LPHAT + 1.96 × $\sqrt{\text{LPHAT} \times (1 - \text{LPHAT}/100)}$"

4. Scroll through the lists. Count the number of intervals that do not contain 0.60, the population proportion. Record the sample proportion for each such interval. What percent of the intervals contain the population proportion p? What percent would you expect to contain p?

5. When analyzing the sample proportions for each confidence interval that did not contain the population proportion, consider the distribution of the sample proportions. The sample proportions are approximately normally distributed with mean 0.60 and standard deviation $\sqrt{0.6(1-0.6)/100} \approx 0.049$. Use **1-Var Stats** to verify that the mean and standard deviation for your one hundred $\hat{p}$ values are consistent with this theory. Notice that each $\hat{p}$ value that did not produce a successful confidence interval is approximately two or more standard deviations away from the mean of 0.6. Calculate the z-scores associated with each unsuccessful $\hat{p}$ value. These z-scores indicate an unusually low or high sample proportion. Therefore, it is not surprising that the confidence intervals they produce do not give accurate information about the population.

Classroom Exercises

1. Graph the normal probability density function, $N(0.6, 0.049)$, that approximates the sampling distribution of the sample proportions in list PHAT. (Define Xmin = .4, Xmax = .8, Ymin = −1, and Ymax = 9.) Press [TRACE] to activate the trace cursor. Then enter a $\hat{p}$-value that resulted in an unsuccessful confidence interval. Notice its location. Repeat for several other similar $\hat{p}$-values. What do you notice about the location of these values?

2. Sort list PHAT in ascending order. (Because lists PLOW and PHIGH are defined using list PHAT, the elements in these lists are moved automatically with respect to the corresponding entry in list PHAT.) Scroll through the lists. Record the least sample proportion and the greatest sample proportion for which the associated confidence interval contained the population proportion p.

3. Do the sample proportions recorded in Exercise 2 produce successful 90% confidence intervals? Do they produce successful 99% confidence intervals? Investigate using **1-PropZInt**. Discuss your results.

4. Determine the width of the 95% confidence intervals created in the activity.

5. Modify the formulas attached to lists PLOW and PHIGH to create 90% confidence intervals. Scroll through the lists to determine how many intervals contain the population proportion p. What percent of the intervals contain the population proportion? (*Hint*: Use **invNorm** to find z^*.)

6. Modify the formulas attached to lists PLOW and PHIGH to create 99% confidence intervals. Scroll through the lists to determine how many intervals contain the population proportion p. What percent of the intervals contain the population proportion? (*Hint*: Use **invNorm** to find z^*.)

7. How wide are the confidence intervals created in Exercises 5 and 6? How does the width of a confidence interval change as the confidence level changes?

Performing a Significance Test for a Population Proportion

In this activity, you can use the TI-83's **DISTR** features to help your students perform a significance test for a population proportion. Then using the TI-83's **1-PropZTest** feature, you can show your students how to perform a significance test for a population proportion automatically. Consider the following problem.

> In a voting precinct, 60% of the voters favor a certain candidate. During the course of a campaign, however, a campaign aide for the candidate becomes concerned that the candidate's support is falling. To determine whether the candidate's support is actually falling, the aide asks the polling staff to poll a simple random sample of 100 voters from the precinct. Only 50% of those polled favor the candidate. Based on this sample, does the campaign aide have cause for concern?

To answer this question, you must determine whether $\hat{p} = 0.50$ is an unusually low sample value given a population proportion of 0.60 and a sample size of 100. If the true population proportion is 0.60, then the sample proportions should be approximately normally distributed with mean 0.6 and standard deviation $\sqrt{0.6(1 - 0.6)/100} \approx 0.049$. What is the probability that a sample proportion from this distribution will be less than or equal to 0.50?

One way to determine this probability is to use the **ShadeNorm** command. Using **ShadeNorm** allows you to determine and graph the area under the normal curve associated with a specified $\hat{p}$ value. For example, to determine the area under the normal curve $N(0.6, 0.049)$ associated with a sample proportion less than or equal to 0.50, do the following.

1. Turn off or clear any existing Y= variables, drawings, and stat plots.

2. Define an appropriate viewing window. In this case, define Xmin = .45, Xmax = .75, Ymin = −2, and Ymax = 8.5.

3. Draw the normal curve and the associated area by executing the following command from the home screen. (See page 25 for more information on **ShadeNorm**.)

    ShadeNorm(0,.5,.6,.049)

A lower bound of 0 is more than 12 standard deviations to the left of the mean. Therefore, very little area is lost by specifying this value rather than negative infinity. Also, recall that executing the command

normalcdf(0,.5,.6,.049)

yields the same area without displaying the graph.

After executing the command, you should obtain a display similar to the one shown below. From the graph, you can see that the area associated with a sample proportion less than or equal to 0.50 is approximately 0.021. Therefore, you can conclude that a $\hat{p}$ value less than or equal to 0.50 is an unusually low value for a population with a true proportion of 0.60. In fact, you can expect such a $\hat{p}$ value only about 2% of the time. The campaign aide does have cause for concern.

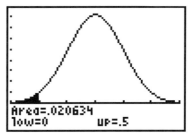

The method described above produces the *P*-value associated with a one-sided significance test for a population proportion. Another way to perform this test is to use the TI-83's **1-PropZTest** feature. To perform a one-sided significance test for a population proportion using **1-PropZTest,** do the following.

1. Press [STAT] [◄] [5] to access the **1-PropZTest** menu.

2. Define the null hypothesis H_0: p_0 = .60, the number of successes: x = 50, the sample size: n = 100, and the alternative hypothesis: prop < p_0.

3. Highlight Draw and press [ENTER]. The resulting display should be similar to the one shown below. Notice that the graph of the normal curve and the corresponding *z*- and *P*-values are displayed. The *P*-value is equal to the area determined using **ShadeNorm.**

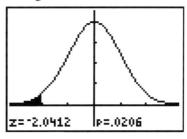

Instead of drawing the graph, you can simply calculate the *P*-value by highlighting Calculate rather than Draw before pressing [ENTER]. The calculator will display the alternative hypothesis, the *z*- and *P*- values, the $\hat{p}$ value, and the sample size.

Inference for Population Proportions

Classroom Exercises

1. Use **1-PropZTest** to determine what sample proportions are significantly lower at the 0.05 level than a population proportion of 0.6 when $n = 100$.

2. Use **1-PropZTest** to investigate how large a sample must be for a sample proportion of 0.50 to be significantly lower at the 0.01 level than a population proportion of 0.60.

3. Use the survey results to create a 95% confidence interval for the true proportion of voters in the precinct who still support the candidate. Express the interval as 0.50 ± margin of error.

4. The margin of error calculated in Exercise 3 is too great to be useful to the campaign aide. The aide prefers a margin of error 0.04 at the 95% confidence level. Assuming another poll produces $\hat{p} = 0.5$, how large a sample is needed to produce a margin of error equal to 0.04? Verify your result using **1-PropZInt**.

5. Suppose the campaign aide is satisfied with a 0.04 margin of error at the 90% confidence level. How large a sample is needed to produce this margin of error? Verify your result using **1-PropZInt**.

20 Comparing Two Proportions

The activities in this chapter describe how to use the TI-83 to help your students investigate and analyze the difference between two proportions. Specifically, the activities in this chapter describe how to use the TI-83 to

- simulate and investigate the sampling distribution of $\hat{p}_1 - \hat{p}_2$.
- create a confidence interval for $p_1 - p_2$.
- perform a significance test for $p_1 - p_2$.

Investigating the Sampling Distribution of $\hat{p}_1 - \hat{p}_2$

In this activity, you will perform simulations to create two lists of sample proportions. Then you will analyze the sampling distribution for the difference of two proportions. Consider the following problem.

> A public university wants to establish an advisory group composed of 100 graduates of the university. The university has thousands of graduates—70% of which are male. A neighboring private university hears of this plan and decides to establish a similar advisory group composed of 100 of its graduates. This university also has thousands of graduates—50% of which are male. If each university chooses the advisory groups by selecting a simple random sample of university graduates, what is a likely difference in the proportion of men in each group?

Simulating Sample Differences $\hat{p}_1 - \hat{p}_2$

To create one difference of sample proportions, use the TI-83's **randBin** command to simulate a sample proportion for each population and then calculate the difference in the proportions. For example, to simulate one difference $\hat{p}_1 - \hat{p}_2$. for the public and private universities, execute the following command from the home screen. (For more information on **randBin,** see page 57.)

$$\underbrace{\text{randBin}(100, .7)/100}_{\hat{p}_1} - \underbrace{\text{randBin}(100, .5)/100}_{\hat{p}_2}$$

To simulate a sampling distribution for $\hat{p}_1 - \hat{p}_2$, you can perform many similar simulations. For example, to create a list containing fifty sample proportion differences $\hat{p}_1 - \hat{p}_2$ for the universities, do the following.

1. Use **SetUpEditor** to restore the standard lists, L1 through L6, in the stat list editor.

2. Define lists L1, L2, and L3 as shown below. Lists L1 and L2 contain fifty sample proportions for the public university and the private university, respectively. List L3 contains the differences of the sample proportions in L1 and L2, $\hat{p}_1 - \hat{p}_2$.

    L1 = randBin(100,.7,50)/100
    L2 = randBin(100,.5,50)/100
    L3 = L1 − L2

    Note: It will take several minutes to perform each simulation.

Analyzing a Sampling Distribution of $\hat{p}_1 - \hat{p}_2$

Summary statistics and graphs are good tools to use when analyzing a sampling distribution of $\hat{p}_1 - \hat{p}_2$.

The mean and standard deviation of the sample proportions in L1 should be approximately 0.7 and $\sqrt{0.7(1 - 0.7)/100} \approx 0.0458$, respectively. The mean and standard deviation of the sample proportions in L2 should be approximately 0.5 and $\sqrt{0.5(1 - 0.5)/100} = 0.05$, respectively. Use **1-Var Stats** to verify that the summary statistics for the sample proportions in L1 and L2 are consistent with these means and standard deviations.

What do you expect the mean of the differences of the sample proportions to equal? What about the standard deviation of these differences? Use **1-Var Stats** to calculate the mean and standard deviation of the differences, $\hat{p}_1 - \hat{p}_2$, in L3. You should observe that the mean difference in the simulated proportions is very close in value to the difference in population proportions. The standard deviation of the differences is larger than the standard deviations for the separate samples but smaller than the sum of these standard deviations.

> Commands **min** and **max** are found under the LIST MATH menu.

To observe the shape of the sampling distribution, define Plot1 to be a histogram with Xlist = L3 and Freq = 1. Set the viewing window by defining Xmin = min(L3), Xmax = max(L3), and Xscl = .03. Define Ymin, Ymax, and Yscl as necessary. You may need to adjust Xscl. The histogram should be fairly symmetric and mound-shaped. Trace the histogram to the bar that contains differences equal to 0.7 − 0.5 = 0.2. Does this appear to be the center of the distribution?

Assessing the Normality of the Sampling Distribution of $\hat{p}_1 - \hat{p}_2$

Because the sample size for each population is large, the distribution of $\hat{p}_1 - \hat{p}_2$ should be approximately normal. To assess the normality of the sample differences in L3, use the 68–95–99.7 Rule as described below.

1. Use the summary statistics, $\bar{x}$ and Sx, associated with the differences in sample proportions to calculate each required interval. The first interval is $\bar{x} \pm Sx$; the second interval is $\bar{x} \pm 2Sx$; the third interval is $\bar{x} \pm 3Sx$.

2. Sort L3 and count the number of observations in each interval.

3. Determine the percent of observations in each interval.

If the percents of observations for the the first interval, the second interval, and the third interval are approximately 68%, 95%, and 99.7%, respectively, you can conclude that the distribution of $\hat{p}_1 - \hat{p}_2$ is approximately normal.

Classroom Exercises

1. Based on your summary statistics for the differences in sample proportions $\hat{p}_1 - \hat{p}_2$, do you expect the public university to select an advisory council with a larger proportion of men than the private university? Explain. What other conclusions can you make from the summary statistics?

2. Use a normal probability plot to assess the normality of the differences in sample proportions. What can you conclude from this graph?

3. Mathematical theory states that the sampling distribution of $\hat{p}_1 - \hat{p}_2$ is approximately normal with mean $\hat{p}_1 - \hat{p}_2$ and standard deviation

$$\sqrt{\frac{p_1(1-p_1)}{n_1} + \frac{p_2(1-p_2)}{n_2}}.$$

Are the results of your simulations consistent with this theory?

Creating a Confidence Interval for $p_1 - p_2$

In this activity, you will learn two methods for using the TI-83 to create a confidence interval for $p_1 - p_2$. Consider the following problem.

> A sociology student wants to determine whether there is a difference between the proportion of female and male students who send birthday cards to their parents. The sociology student selects a simple random sample of 60 female students and 60 male students. Each is asked if he or she sent a birthday card to a parent during the last twelve months. Forty-two females and thirty-six males reply "yes". Create a 90% confidence interval for the difference between the true population proportions.

One way to create a confidence interval for the difference between the population proportions, $p_1 - p_2$, is to use the definition of confidence interval. By definition, the confidence interval for $p_1 - p_2$ is

$$(\hat{p}_1 - \hat{p}_2) \pm z^* \sqrt{\frac{\hat{p}_1(1-\hat{p}_1)}{n_1} + \frac{\hat{p}_2(1-\hat{p}_2)}{n_2}}$$

where n_1 and n_2 are the sample sizes and z^* is the appropriate standard normal value. Note that the confidence interval is not valid for small sample sizes.

To create a 90% confidence interval for the difference between the male and female population proportions, do the following.

1. Determine $\hat{p}_1$ and $\hat{p}_2$.

$$\hat{p}_1 = \frac{42}{60} = 0.7 \qquad \hat{p}_2 = \frac{36}{60} = 0.6$$

2. Determine z^*. The z^*-value corresponding to a 90% confidence is approximately 1.645. You can calculate z^* using the command **invNorm(.95)**. (See page 29 for more information on **invNorm**.)

3. Determine the confidence interval by calculating the left and right endpoints. The left and right endpoints are

$$(\hat{p}_1 - \hat{p}_2) \pm z^* \sqrt{\frac{\hat{p}_1(1-\hat{p}_1)}{n_1} + \frac{\hat{p}_2(1-\hat{p}_2)}{n_2}}$$

$$\approx (0.7 - 0.6) \pm 1.645 \sqrt{\frac{0.7(1-0.7)}{60} + \frac{0.6(1-0.6)}{60}}$$

$$\approx 0.1 \pm 0.142$$

Therefore, the confidence interval is $(-0.042, 0.242)$. Note that this confidence interval contains the value 0, which implies that there may be no difference between the true population proportions.

Another way to create a confidence interval for the difference between two population proportions is to use the TI-83's **2-PropZInt** feature. Using this feature allows you to create a confidence interval automatically. For example, to create a 90% confidence interval for the difference between the male and female population proportions, do the following.

1. Press [STAT] [◄] [ALPHA] [B] to access the **2-PropZInt** menu.

2. The **2-PropZInt** menu prompts you to enter five values: the number of successes for each sample, the sample size of each sample, and the confidence level. For this problem, define x1 = 42, n1 = 60, x2 = 36, n2 = 60, and C-Level = .90.

3. Highlight Calculate and press [ENTER]. The calculator will display the confidence interval $(-.0424, .24245)$, $\hat{p}_1 = 0.7$, $\hat{p}_2 = 0.6$, $n_1 = 60$, and $n_2 = 60$.

Classroom Exercises

1. Calculate the width of the 90% confidence interval. How does the width of the interval change if the sample sizes increase to 100 and the sample proportions remain the same? What if the sample sizes increase to 200? Investigate using **2-PropZInt.** Discuss your results.

2. Assume you select equal sample sizes from both populations of students. How large must the sample sizes be in order for the 90% confidence interval to no longer contain 0?

3. How does the width of the confidence interval change as the level of confidence increases? How large must the sample sizes be to produce a 99% confidence interval that does not contain 0?

Performing a Significance Test for $p_1 - p_2$

In this activity, you will learn two methods for using a TI-83 to perform a significance test for $p_1 - p_2$. Consider the following problem.

> A sociology student, who attends a college where the women's soccer team is nationally ranked, wants to determine whether there is a difference in the proportion of female students and the proportion of male students who attend the women's home soccer games. Halfway through the soccer season, the sociology student selects a random sample of 100 female students and 100 male students and asks whether they have attended a women's home soccer game during the current season. Twenty-two females and thirty males

Comparing Two Proportions   127

report that they have attended at least one game. Based on these samples, should the sociology student conclude that there is a difference between the true population proportions?

To solve this problem, you must determine whether the observed difference of 8% between the sample proportions is evidence of a real difference in the population proportions or is likely to be the result of sampling variability. Because the student has no idea which true population proportion may be higher, you should test the null hypothesis that there is no difference between the proportions against the two-sided alternative hypothesis that there is a difference.

You should use the z statistic as the test statistic because the sampling distribution of $\hat{p}_1 - \hat{p}_2$ is approximately normal. The null hypothesis states that the two population proportions are equal. Therefore, instead of using $\hat{p}_1$ and $\hat{p}_2$ as estimates for the individual population proportions in the formula for the standard error of $\hat{p}_1 - \hat{p}_2$, you can use a pooled sample proportion as the estimate for the single proportion that applies to both populations if H_0 is true. The pooled sample proportion is simply the ratio of total successes from both samples to the combined sample size. In this example, the pooled sample proportion, $\hat{p}$, is

$$\frac{(30 + 22)}{(100 + 100)} = 0.26.$$

Therefore, the z statistic is

$$z = \frac{(\hat{p}_1 - \hat{p}_2)}{\sqrt{\hat{p}(1-\hat{p})\left(\frac{1}{n_1} + \frac{1}{n_2}\right)}} = \frac{(0.3 - 0.22)}{\sqrt{0.26(1-0.26)\left(\frac{1}{100} + \frac{1}{100}\right)}} \approx 1.2897.$$

Because the alternative hypothesis in this example is two-sided, the P-value for this test is the sum of the area to the left of $z = -1.2897$ and the area to the right of $z = 1.2897$. To calculate this area, execute the command

$$1 - \text{normalcdf}(-1.2897, 1.2897)$$

from the home screen. (For more information on **normalcdf**, see page 28.) The result is approximately 0.197. A P-value of 0.197 does not indicate strong evidence against the null hypothesis. Therefore, you should conclude that the observed difference in the sample proportions could result from sampling variability.

The method described above produces the P-value associated with a two-sided significance test for the difference in two population proportions. Another way to perform this test is to use the TI-83's **2-PropZTest** feature. To perform a two-sided significance test for $p_1 - p_2$ using **2-PropZTest,** do the following.

1. Press ⎡STAT⎤ ◂ ⎡6⎤ to access the **2-PropZTest** menu.

2. The **2-PropZTest** menu prompts you to enter four values: the number of successes for each sample and the sample size of each sample. The menu also prompts you to specify the alternative hypothesis. For this example, define x1 = 30, n1 = 100, x2 = 22, n2 = 100, and p1 ≠ p2 for the alternative hypothesis.

3. Highlight Draw and press ⎡ENTER⎤. The resulting display should be similar to the one shown below. The graph of the normal curve and the corresponding z- and P-values are displayed.

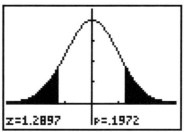

Instead of drawing the graph, you can simply calculate the P-value by highlighting Calculate rather than Draw before pressing ⎡ENTER⎤. The calculator will display the alternative hypothesis, the z- and P-values, the sample proportions $\hat{p}_1$ and $\hat{p}_2$, the pooled sample proportion $\hat{p}$, and the sample sizes from each population, n_1 and n_2.

Classroom Exercises

1. Use the sample results to create a 90% confidence interval for the difference between the proportion of female and male students who attend women's home soccer games. Do you expect this interval to contain 0? Explain.

2. Suppose the sample proportions remain unchanged but both sample sizes increase to 200. Is the difference in proportions, 8%, significant at the $\alpha = 0.05$ level? If not, how large must the sample sizes (assume equal sample sizes) be in order for the observed difference to be statistically significant at this level? Investigate using **2-PropZTest.**

3. Determine whether the proportion of females who send birthday cards to parents is significantly higher at the $\alpha = 0.05$ level than the proportion of males who send birthday cards to parents. Use **2-PropZTest.**

4. Perform a simulation to estimate the *P*-value associated with the significance test for this activity. Based on information from the samples, if the null hypothesis is true and the population proportions are equal, you would assume $p = 0.26$. The *P*-value is the probability that the sample proportions of 100 male students and 100 female students would differ by as much or more than the observed difference of 0.08 if the samples are selected from populations where the true proportion is 0.26. Define lists L1, and L2, and L3 as shown below.

$$L1 = \text{randBin}(100,.26,50)/100$$
$$L2 = \text{randBin}(100,.26,50)/100$$
$$L3 = L1 - L2$$

L1 contains simulated proportions of males who had attended a women's soccer game in 50 samples of size 100. L2 contains corresponding sample proportions for females. To estimate the *P*-value, determine the percent of differences in L3 that are less than or equal to -0.08 or greater than or equal to 0.08. To obtain a better estimate for the *P*-value, pool the simulation results of all students in the class.

21 Chi-Square Tests

The activities in this chapter describe how to use the TI-83 to help your students better understand the chi-square distributions and various chi-square tests. Specifically, the activities in this chapter show how to use the TI-83 to

- graph density curves for chi-square distributions.
- calculate areas associated with chi-square distributions.
- perform a chi-square test for goodness of fit.
- perform a chi-square test for independence.

Graphing Density Curves for Chi-Square Distributions

In this activity, you will use the TI-83 to graph chi-square distributions with various degrees of freedom. Consider the following problem.

> Graph the density curves for the chi-square distributions with 1, 3, and 10 degrees of freedom. Compare the graphs.

To graph the density curve for each distribution using the TI-83, do the following.

1. Turn off or clear any existing Y= variables, stat plots, and drawings.

2. Set an appropriate viewing window for the density curves. In this case, define Xmin = 0, Xmax = 15, Xscl = 1, Ymin = 0, Ymax = .5, and Yscl = .1.

3. Access the Y= editor and define **Y1 = X^2pdf(X,1).** Access command X^2**pdf** by pressing [2nd] [DISTR] [6]. When using X^2**pdf,** you must specify the degrees of freedom. Press [GRAPH] to display the density curve for the chi-square distribution with 1 degree of freedom. The calculator display should be similar to the one shown on the next page. Notice that the curve is very different from the normal and $t(k)$ density curves. The density curve for the chi-square distribution with 1 degree of freedom is not symmetric and is strongly skewed to the right.

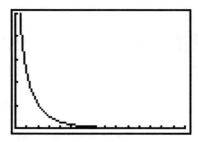

4. Access the Y= editor and define **Y2 = X^2pdf(X,3)**. Press GRAPH to display the density curves for the chi-square distributions with 1 and 3 degrees of freedom. The calculator display should be similar to the one shown below. Notice that the density curve for the chi-distribution with 3 degrees of freedom also lacks symmetry and is skewed to the right, but not as much as the density curve with 1 degree of freedom.

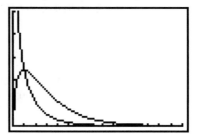

5. Access the Y= editor and define **Y3 = X^2pdf(X,10)**. Press GRAPH to display the density curves for the chi-square distributions with 1, 3, and 10 degrees of freedom. The calculator display should be similar to the one shown below. Notice that the density curve for the chi-square distribution with 10 degrees of freedom is flatter and more symmetric than the other density curves. It is also less skewed.

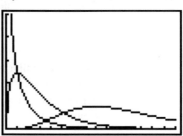

Classroom Exercises

1. By definition, the chi-square statistic is always greater than or equal to zero. Change the window settings to verify that the graphs of the chi-square distributions with 1, 3, and 10 degrees of freedom are defined for positive values only.

2. Graph the density curve for the chi-square distribution with 25 degrees of freedom. Change the window settings as needed. How does the resulting graph compare to the density curves for the chi-square distributions with 1, 3, and 10 degrees of freedom? As the degrees of freedom increase, how does the graph of the chi-square distribution change?

Calculating Areas Associated with Chi-Square Distributions

In this activity, you will learn two methods for using the TI-83 to calculate the area under density curves for chi-square distributions with various degrees of freedom. Consider the following problem.

> Calculate the area to the right of $X^2 = 5$ for chi-square distributions with 1, 3, and 10 degrees of freedom.

One way to determine areas associated with chi-square distributions is to use the TI-83's **ShadeX²** command. To access **ShadeX²**, enter the following keystrokes.

$\boxed{\text{2nd}}$ [DISTR] $\boxed{\blacktriangleright}$ $\boxed{3}$ *Access **ShadeX²** command.*

When using **ShadeX²**, you must define the area to be calculated by specifying the lower and upper bounds and the degrees of freedom of the chi-square distribution. For example, to use **ShadeX²** to calculate the area to the right of $X^2 = 5$ for the chi-square distributions with 1, 3, and 10 degrees of freedom, do the following.

1. Turn off or clear any existing Y= variables, stat plots, and drawings.

2. Define Xmin = 0, Xmax = 15, Xscl = 1, Ymin = −.1, and Ymax = .3. It is important that Ymin = −.1 so that the graph captions do not interfere with the graphs.

3. Calculate the area to the right of $X^2 = 5$ for the chi-square distribution with 1 degree of freedom by executing **ShadeX²(5,1E99,1)** from the home screen. The values 5, 1E99, and 1 specify the lower bound, the upper bound, and the degrees of freedom, respectively. The calculator display should be similar to the one shown on the next page. Notice that

the area to the right of $X^2 = 5$ for the chi-square distribution with 1 degree of freedom is barely visible using the present window settings. The area, however, is approximately 0.025347. So for a chi-square distribution with 1 degree of freedom, a chi-square value of 5 is very high.

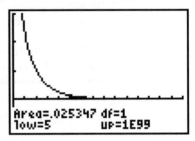

4. Use **ClrDraw** to clear the existing drawing. Then, in a similar manner, calculate the area to the right of $X^2 = 5$ for the chi-square distributions with 3 and 10 degrees of freedom by executing **ShadeX²(5,1E99,3)** and **ShadeX²(5,1E99,10)** from the home screen. The resulting calculator displays for these commands should look similar to the ones shown below. Notice that the area to the right of $X^2 = 5$ for the chi-square distribution with 3 degrees of freedom is approximately 0.171797 and the area to the right of $X^2 = 5$ for the chi-square distribution with 10 degrees of freedom is approximately 0.891178. For a chi-square distribution with 10 degrees of freedom, a chi-square value of 5 is relatively low.

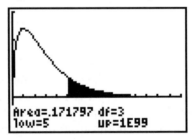 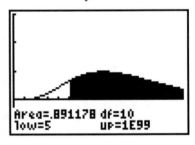

Another way to determine areas associated with chi-square distributions is to use the TI-83's X^2**cdf** command. To access X^2**cdf**, enter the following keystrokes.

[2nd] [DISTR] [7]   *Access X^2cdf command.*

When using X^2**cdf,** you must define the area to be calculated by specifying the lower and upper bounds. You must also specify the degrees of freedom of the chi-square distribution. For example, to use X^2**cdf** to calculate the area to the right of $X^2 = 5$ for the chi-square distributions with 1, 3, and 10 degrees of freedom, execute the commands on the next page.

X^2cdf(5,1E99,1)
X^2cdf(5,1E99,3)
X^2cdf(5,1E99,10)

The results should be equal to those obtained when calculating the areas graphically.

Classroom Exercises

1. Use the methods discussed in the activity to calculate the area to the right of $X^2 = 10$ for the chi-square distributions with 1, 3, and 10 degrees of freedom. How does the area change as the degrees of freedom increase?

2. Verify that the total area under the density curve of the chi-square distribution with 1 degree of freedom is 1.

Performing a Chi-Square Test for Goodness of Fit

Performing a goodness of fit test helps determine whether a population has a specified theoretical distribution. For example, a chi-square test for goodness of fit measures, using the chi-square statistic, X^2, how well a sample distribution fits the theoretical distribution. The formula for the chi-square statistic, X^2, is

$$X^2 = \sum_{k=1}^{n} \frac{(\text{obs}_k - \text{exp}_k)^2}{\text{exp}_k}$$

where obs_k represents the observed frequencies in a sample, exp_k represents the corresponding expected frequencies based on the hypothesized distribution, and n represents the number of outcome categories, or cells. To use the chi-square test for goodness of fit, all expected frequencies should be greater than or equal to 5.

The sampling distribution of the X^2 statistic is closely approximated by the chi-square distribution with $(n - 1)$ degrees of freedom. Small X^2 values occur when the observed and expected frequencies are approximately equal. Therefore, small X^2 values support the hypothesis that the population from which the sample is drawn is distributed according to the theoretical distribution. Similarly, large X^2 values occur when the fit is not good, that is, the differences between the observed and expected frequencies are large. Therefore, large X^2 values support rejection of the hypothesis that the sample is drawn from a population that is distributed as hypothesized.

In this activity, you will learn how to perform a chi-square goodness of fit test using the TI-83. Consider the following problem.

> Suppose you construct a spinner for use at a school carnival. The spinner is attached to a circular board that is partitioned into five equal sectors. Each sector is painted blue, red, yellow, green, or white. To determine whether the spinner is balanced properly, you spin the spinner 100 times and record the color the spinner is pointing to when it stops. You obtain the following results.
>
> Blue: 16   Red: 22   Yellow: 26
> Green: 17   White: 19
>
> Are these results consistent with the expected results from a balanced spinner?

You can expect a fair spinner with uniform partitioning to have a uniform distribution of outcomes; therefore, you should test the hypothesis that the distribution of outcomes is 20% blue, 20% red, 20% yellow, 20% green, and 20% white. Theoretically, spinning the spinner 100 times should result in 20 occurrences of each color.

The TI-83 does not have a built-in statistical test that computes a chi-square statistic to test for goodness of fit. You can, however, perform a chi-square test for goodness of fit using the TI-83's list features. For example, to perform a chi-square goodness of fit test to help determine whether the spinner is balanced, do the following.

1. Use **SetUpEditor** to restore the standard lists, L1 through L6, to the stat list editor. Clear L1, L2, and L3.

2. Enter the observed and expected frequencies in lists L1 and L2, respectively.

    L1 = {16, 22, 26, 17, 19}   L2 = {20, 20, 20, 20, 20}

3. Define L3 = (L1 − L2)²/L2.

4. Determine the chi-square statistic by finding the sum of L3: **sum(L3)**. The result is 3.3.

5. Determine the area to the right of $X^2 = 3.3$ under the density curve of the chi-square distribution with $(n - 1) = 4$ degrees of freedom. To calculate this area, execute the command X^2**cdf(3.3,1E99,4)**. The resulting area, or *P*-value, is approximately 0.5089. Therefore, you can conclude that about 51% of the time you spin a fair spinner 100 times, the results will vary, by chance, from uniform as much or more than observed. The observed results are consistent with those expected from a balanced spinner.

Classroom Exercises

1. To simulate the frequency of one color when spinning a fair spinner 100 times, execute the command **randBin(100,.2)**. Discuss the results of executing the following commands. (Assume L2 = {20, 20, 20, 20, 20}.)

    seq(randBin(100,.2), X, 1, 4, 1) → L1:100 − sum(L1) → L1(5)
    :(L1−L2)²/L2 → L3: sum(L3)

2. Use the simulation commands in Exercise 1 to simulate ten values for the chi-square statistic. How does the chi-square statistic obtained in the activity, $X^2 = 3.3$, compare to the simulated X^2 values? What percent of the simulated X^2 values are greater than or equal to $X^2 = 3.3$? Note that the resulting percent approximates the P-value for $X^2 = 3.3$ with 4 degrees of freedom. To better approximate the P-value, pool the results of all students in the class.

3. Use X^2**cdf** to calculate the P-value associated with the largest X^2 value simulated in Exercise 2. Would the resulting P-value lead to a correct conclusion or to a Type I error? Explain.

4. Perform a chi-square goodness of fit test to determine whether the TI-83's random integer feature is working properly.

    a. Use **randInt** to generate 100 integers between 0 and 9. Store the integers in L4.
    b. Use Plot1 to create a histogram that represents the integers in L4. Define Xmin = −.5, Xmax = 9.5, Xscl = 1, Ymin = −5, Ymax = 20, and Yscl = 5.
    c. Trace the histogram and record the frequency of each integer in list L1.
    d. Store the corresponding expected frequencies in list L2.
    e. Calculate the chi-square statistic and the corresponding P-value as described in the activity.

    Discuss your results.

Performing a Chi-Square Test for Independence

Another chi-square test, the chi-square test for independence, is used to test the hypothesis of independence between two categorical variables. In this activity, you will learn how to use the TI-83 to perform a chi-square test for independence. Consider the following problem.

> The student newspaper staff at a high school conducted a poll to determine the student body's opinion regarding the school's disciplinary policy. In a simple random sample, 266 students were asked if they thought the current disciplinary policy was effective. The polled students were directed to

answer yes, no, or undecided. They were also asked to specify their grade level: freshman, sophomore, junior, or senior. The results of the survey are listed below.

	Freshman	Sophomore	Junior	Senior
Yes	22	10	15	8
No	31	28	60	47
Undecided	15	11	12	7

Based on these results, can the newspaper staff conclude that opinions regarding the school's disciplinary policy are independent of grade level?

Using the TI-83's X^2-**Test** feature, you can automatically perform a chi-square test for independence. The X^2-**Test** feature allows you to determine the expected frequencies, the chi-square statistic, and the P-value based on a set of observed frequencies that are stored in a matrix. For example, to perform a chi-square test for independence using the newspaper staff's survey results, do the following.

1. Store the observed frequencies in a matrix. To store the newspaper staff's survey results in matrix A, enter the following keystrokes.

MATRX ◄ 1	*Access matrix A.*
3 ENTER 4 ENTER	*Specify row and column dimensions.*
22 ENTER 10 ENTER	*Enter observed frequencies.*
15 ENTER 8 ENTER	
31 ENTER 28 ENTER	
60 ENTER 47 ENTER	
15 ENTER 11 ENTER	
12 ENTER 7 ENTER	

2. Press STAT ◄ ALPHA [C] to access the X^2-**Test** menu.

3. The X^2-**Test** menu prompts you to specify the matrix containing the observed frequencies and the matrix in which you want to store the expected frequencies. For this example, define Observed = [A] and Expected = [B]. To specify a matrix name, simply paste the appropriate matrix name from the matrix names menu.

4. Highlight Draw and press ENTER. The resulting display should be similar to the one shown on the next page. Notice that $X^2 \approx 16.0884$ and $P \approx 0.0133$. Because the X^2 value is so large, the shaded region under the chi-square density curve is not visible. Also, because the resulting P-value is so small, the student newspaper staff should reject the hypothesis that opinions regarding the school's disciplinary policy are independent of grade level.

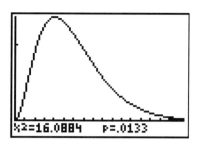

Instead of drawing the graph to calculate X^2 and P, highlight Calculate rather than Draw before pressing ENTER. The calculator will display the degrees of freedom ($df = 6$) in addition to the X^2 statistic and the P-value.

To gain information about what causes this X^2 statistic to be so large, compare the matrix of expected frequencies to the observed frequencies. To display the expected frequencies on the home screen, press MATRX [2] ENTER. Use the cursor arrows to scroll through the matrix.

Classroom Exercises

1. Use the method for performing a chi-square test for independence discussed in your statistics textbook and the TI-83's X^2**cdf** feature to verify the results obtained by the TI-83.

2. If you completed the activities in Chapter 20, recall that you performed a test to compare the proportion of males and females who attended women's home soccer games. The test was based on samples of 100 female students and 100 male students. Perform a chi-square test to compare these proportions by entering the following observed frequencies into matrix A.

	Females	Males
Yes	22	30
No	78	70

   Use the output of the X^2-**Test** and **2-PropZTest** to verify that the X^2 statistic equals the square of the z statistic and that the P-values obtained for both tests are equal.

   The test in Exercise 2 is an example of a chi-square test for homogeneity. In this situation, the column totals are predetermined and fixed by the decision to sample 100 males and 100 females.

3. A high school teacher wants to determine whether there is a difference in the proportion of 10th, 11th, and 12th grade students who have part-time jobs. The teacher surveys a random sample of fifty students in each of these grades and asks them if they have a part-time job. Twelve 10th grade students, seventeen 11th grade students, and twenty 12th grade students responded "yes". Perform a chi-square test to determine whether the proportion of 10th, 11th, and 12th grade students who have a part-time job differs.

22 Inference for Slope of a Regression Line

In Chapter 8, you learned how to use the TI-83 to create a scatter plot, calculate the equation of a least-squares regression line $\hat{y} = a + bx$, calculate the correlation coefficient r, and calculate and graph residuals associated with the least-squares regression line. The activities in this chapter extend the topic of least-squares regression lines. Specifically, the activities in this chapter describe how to use the TI-83 to

- calculate standard error about a regression line.
- perform a significance test for slope.
- create a confidence interval for slope.

Before beginning the activities, you should discuss with your students when it is appropriate to perform inference procedures for the slope of a regression line. For example, such inference procedures should not be performed if the data set contains any outliers or influential points. Also, for the inference procedures to be valid, the following assumptions must be satisfied.

1. For any fixed value of the explanatory variable, x, the values of the response variable, y, are independent of each other and are normally distributed.

2. The true relationship between x and the mean response μ_y can be modeled by a linear function with parameters α (the y-intercept) and β (the slope). These parameters are estimated by the slope and y-intercept of the least-squares regression line.

3. For any value x, the standard deviation of the response variable y is σ. The standard deviation of the responses about the true regression line, σ, is estimated by the standard error about the line. For a sample of size n, the formula for the standard error about a regression line is

$$s = \sqrt{\frac{\Sigma \, (\text{residuals})^2}{n-2}} = \sqrt{\frac{\Sigma (y_i - \hat{y}_i)^2}{n-2}}.$$

Calculating Standard Error About a Regression Line

In this activity, you will use the TI-83 to verify whether the inference procedures for the slope of a regression line are valid for a given data set. Then after verifying the validity of the inference procedures, you will calculate the standard error about the regression line. Consider the following problem.

A student is studying the relationship between the heights of fathers and their sons. The heights, in centimeters, of fifteen father/son pairs between the ages of 20 and 55 are listed below.

Father's Height	Son's Height
172	167
171	176
178	172
167	163
174	170
185	184
170	173
174	172
180	183
176	175
182	181
176	175
178	179
176	174
179	172

Table 22.1

Verify that the inference procedures for the slope of a regression line are valid for this data set and find the equation of the least-squares regression line for the data. Then calculate the standard error about the regression line.

Verifying Validity of Inference Procedures

To verify that the inference procedures for the slope of a regression line are valid for a particular data set, you should graph the data using a scatter plot, calculate a least-squares regression line for the data, and examine the residuals. For example, to use the TI-83 to verify that the inference procedures for the slope of a regression line are valid for the father/son height data, do the following.

1. Use **SetUpEditor** to create two lists named FATHR and SON. Then enter the data into these lists.

2. Use the TI-83 to graph the scatter plot and least-squares regression line for the data. Graph the scatter plot so that father's height is the explanatory variable and son's height is the response variable. The resulting scatter plot and regression line should be similar to the graph shown below. Notice that the scatter plot indicates a positive linear relationship between the father's and son's heights and that there appear to be no outliers or other influential points. The equation of the least-squares regression line is $\hat{y} = 5.1803 + 0.9622x$ with correlation coefficient $r \approx 0.7993$.

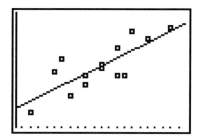

3. Recall that the TI-83 automatically creates a list of residuals, RESID, whenever it calculates the equation of a least-squares regression line. Copy the contents of list RESID to a new list named RESFS. Then use lists FATHR and RESFS to create a residual plot. The residual plot should look similar to the one shown below. Notice that there does not seem to be any apparent pattern. The lack of an apparent pattern confirms a linear relationship between the variables. Because the magnitude of the residuals does not appear to change significantly as the explanatory variable changes, you can assume that the standard deviation of the response variable is constant.

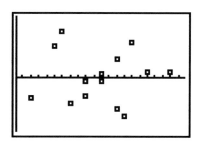

4. Create a histogram or normal probability plot of the residuals to determine whether the distribution of residuals is approximately normal. Checking for normally distributed residuals is an indirect check that the values of the response variable vary according to a normal distribution for any fixed value of the explanatory variable. A histogram of the residuals in list RESFS should look similar to the one shown below. The histogram is fairly symmetric and mound-shaped which is consistent with a normal distribution.

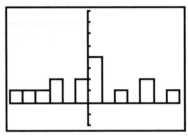

Based on the results of the scatter plot, the residual plot, and the histogram, you can conclude that it is valid to perform inference procedures for the slope of the regression line given by $\hat{y} = 5.1803 + 0.9622x$.

Calculating Standard Error

To calculate the standard error about the regression line $\hat{y} = 5.1803 + 0.9622x$, do the following.

1. Execute the command **1-Var Stats LRESFS** to calculate the sum of the squared residuals. You should obtain $\Sigma x^2 = 165.282306$.

2. Execute the command $\sqrt{(\Sigma x^2 / 13)}$. The result, the standard error s about the regression line, is approximately 3.5657.

> Press [VARS] [5] [▶] [2] to access Σx^2.

Performing a Significance Test for β

In this activity, you will learn how to use a TI-83 to perform a significance test for the slope of a regression line. Consider the following problem.

> Using the father/son height data provided in the previous activity and the corresponding least-squares regression line $\hat{y} = 5.1803 + 0.9622x$, perform a significance test to determine whether the slope of the regression line is significantly greater than 0. That is, test the null hypothesis $H_0: \beta = 0$ against the alternative hypothesis $H_a: \beta > 0$.

One way to perform this significance test is to compute the t statistic

$$t = \frac{b}{SE_b}$$

where b is the slope of the least-squares regression line and SE_b is the standard error of the least-squares slope. The formula for SE_b is

$$SE_b = \frac{s}{\sqrt{\Sigma(x_i - \bar{x})^2}}$$

where s is the standard error about the regression line and $\Sigma(x_i - \bar{x})^2$ is the sum of the squared differences between the values of the explanatory variable and the mean of the explanatory variable. The degrees of freedom for the t statistic are $n - 2$, where n is the sample size.

Using these formulas and the value $s = 3.5657$, you can calculate the t statistic as shown below. Round-off error is expected if exact values are not used in the calculations.

> The standard deviation of list FATHR is calculated using the formula
> $$\sqrt{\frac{\Sigma(x_i - \bar{x})^2}{(n-1)}}.$$
> Therefore, an efficient way to evaluate $\sqrt{\Sigma(x_i - \bar{x})^2}$ is to use the TI-83 to calculate the standard deviation of list FATHR and multiply the result by $\sqrt{14}$. The result is approximately 17.7689.

$$t = \frac{b}{SE_b} = \frac{b}{\left(\frac{s}{\sqrt{\Sigma(x_i - \bar{x})^2}}\right)} \approx \frac{0.9622}{\left(\frac{3.5657}{17.7689}\right)} \approx \frac{0.9622}{0.2007} \approx 4.794$$

Note that SE_b is approximately 0.2007.

The P-value associated with this test is equal to the area to the right of $t = 4.794$ under the $t(13)$ density curve. To calculate this area, execute the command

tcdf(4.794,1E99,13)

from the home screen. The result is approximately 0.000175 which provides strong evidence that the null hypothesis of $\beta = 0$ should be rejected.

The method described above produces the P-value associated with a one-sided significance test for the slope of a regression line. Another way to perform this test is to use the TI-83's **LinRegTTest** feature. To perform a one-sided significance test for the slope of a regression line using **LinRegTTest,** do the following.

Inference for Slope of a Regression Line

Entering a Y= variable after RegEQ is optional. If you do enter a Y= variable, the regression equation is stored in that variable when **LinRegTTest** is executed.

1. Press [STAT] [◄] [ALPHA] [E] to access the **LinRegTTest** menu.

2. The **LinRegTTest** menu prompts you to enter the lists in which the data are stored, the frequency of the data, the alternative hypothesis, and a location for the regression equation. For this example, define Xlist = FATHR, Ylist = SON, Freq = 1, $\beta \ \& \ \rho > 0$ for the alternative hypothesis, and RegEQ = Y1.

3. Highlight Calculate and press [ENTER]. The calculator will display the alternative hypothesis, the t statistic $t \approx 4.795$, the P-value $P \approx 0.000175$, the degrees of freedom $df = 13$, the slope and y-intercept of the regression line $b \approx 0.9622$ and $a \approx 5.1803$, the standard error about the regression line $s \approx 3.5657$, the correlation coefficient $r \approx 0.799$, and the coefficient of determination $r^2 \approx 0.639$.

Creating a Confidence Interval for β

In this activity, you will use the TI-83 to create a confidence interval for the slope β of the true regression line. Consider the following problem.

> Using the father/son height data provided in the previous activities and the corresponding least-squares regression line $\hat{y} = 5.1803 + 0.9622x$, create a 95% confidence interval for the slope β of the true regression line relating the father/son heights.

To create a confidence interval for β, use the definition of confidence interval. By definition, the confidence interval for the slope β of the true regression line is

$$b \pm t^* SE_b$$

where b is the slope of the least-squares regression line and SE_b is the standard error of the least-squares slope. The t^* value is based on the t distribution with $(n - 2)$ degrees of freedom.

To create a 95% confidence interval for the slope β of the true regression line for the father/son data, do the following.

1. Determine b and SE_b. From the results of previous activities in this chapter, $b \approx 0.9622$ and $SE_b \approx 0.2007$.

2. Use the TI-83's equation solver to determine t^*. In this example, use an initial guess of 2 to solve the equation **0 = tcdf(−T,T,13)−.95**. You should obtain $t^* \approx 2.1604$.

3. Determine the confidence interval by calculating the left and right endpoints. The left and right endpoints are

$$b \pm t^*SE_b \approx 0.9622 \pm 2.1604(0.2007)$$
$$\approx 0.9622 \pm 0.4336.$$

The confidence interval is (0.5286, 1.3958). Therefore, you can be 95% confident that an increase of 1 centimeter in the father's height is associated with an increase of between 0.5 and 1.4 centimeters in the son's height.

Classroom Exercises

1. By definition, the test statistic t is $t = b/SE_b$, where b is the slope of the least-squares regression line and SE_b is the standard error of the least-squares slope. Solving this equation for SE_b yields $SE_b = b/t$. Use the output of **LinRegTTest** to verify that $SE_b \approx 0.2007$.

2. Use **1-Var Stats** to calculate the standard deviation of the heights of the fifteen sons. Compare the result to the standard error about the regression line. What can you conclude?

3. Number the father/son pairs in the order they are listed in the original data set. Then use the **randInt** command to randomly select five pairs of data and delete them from the data set. Calculate the least-squares regression line for the revised data. Discuss the effects on the regression line equation and the values of r and r^2.

4. Using the revised data from Exercise 3, perform a significance test to determine whether the positive association between the father/son heights is significant.

5. Using the revised data from Exercise 3, create a 95% confidence interval for the slope β of the true regression line. Compare your results to those obtained in the activity.

23 One-Way Analysis of Variance

Analysis of variance (**ANOVA**) tests are often used to determine whether there are differences between two or more population means. To perform analysis of variance tests, the data being analyzed should be independent simple random samples from normal populations, and each population should have the same standard deviation. One-way analysis of variance tests use the F statistic.

In this chapter, you will learn how to use the TI-83 to

- graph density curves for F distributions.
- perform one-way **ANOVA** tests.
- create a confidence interval for a population mean μ_i.

Graphing Density Curves for F Distributions

The F distributions are a family of distributions with two parameters: the degrees of freedom for the numerator, and the degrees of freedom for the denominator. In this activity, you will use the TI-83 to graph several F distributions. Consider the following problem.

> Graph the density curves for the $F(2, 28)$, $F(4, 50)$, and $F(10, 100)$ distributions.

To graph the density curve for each distribution using the TI-83, do the following.

1. Turn off or clear any existing Y= variables, stat plots, and drawings.

2. Set an appropriate viewing window for the density curves. In this case, define Xmin = 0, Xmax = 5, Xscl = 1, Ymin = −.1, Ymax = 1, and Yscl = .1.

3. Access the Y= editor and define **Y1 = Fpdf(X,2,28)**. You can access command **Fpdf** by pressing [2nd] [DISTR] [8]. When using **Fpdf**, you must specify the degrees of freedom in the numerator and the denominator. Press [GRAPH] to display the density curve for the F distribution with 2 degrees of freedom in the numerator and 28 degrees of freedom in the denominator. The calculator display should be similar to the one shown on the next page. Notice that the curve is very different from the normal and $t(k)$ density curves.

The density curve for the $F(2, 28)$ distribution is not symmetric and is strongly skewed to the right.

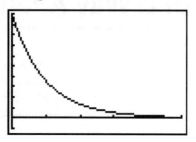

4. Access the Y= editor and define **Y2 = Fpdf(X,4,50).** Press GRAPH to display the density curves for the $F(2, 28)$ and $F(4, 50)$ distributions. The calculator display should be similar to the one shown below. Notice that the density curve for the $F(4, 50)$ distribution also lacks symmetry and is skewed to the right, but not as much as the density curve for the $F(2, 28)$ distribution.

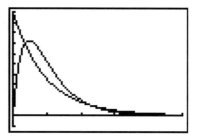

5. Access the Y= editor and define **Y3 = Fpdf(X,10,100).** Press GRAPH to display the density curves for the $F(2, 28)$, $F(4, 50)$ and $F(10, 100)$ distributions. The calculator display should be similar to the one shown below. Notice that the density curve for the $F(10, 100)$ distribution is more symmetric and is also less skewed to the right than the other density curves.

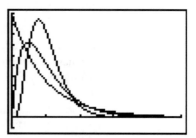

Performing a One-Way ANOVA Test

In this activity, you will use the TI-83 to perform a one-way **ANOVA** test. Consider the following problem.

> Your friend makes three paper airplanes, each a different model, and wants to determine whether there are differences in the mean length of time each model stays airborne. To help your friend analyze the lengths of flight times, you launch each model ten times and time the flight. (Assume ideal conditions for the flight launches: no arm fatigue, no wear and tear on the planes, alternating models, etc.) The lengths of flight times, in seconds, for each plane model are listed below.
>
> Plane A: 1.22, 1.35, 1.26, 1.31, 1.25, 1.16, 1.20, 1.08, 1.42, 1.13
> Plane B: 1.19, 1.33, 1.43, 1.49, 1.14, 1.24, 1.26, 1.31, 1.40, 1.36
> Plane C: 1.20, 1.58, 1.55, 1.26, 1.47, 1.64, 1.34, 1.51, 1.31, 1.41
>
> Based on these data, does it appear that there is a difference in the mean length of time each model stays airborne?

Using the TI-83's **ANOVA** feature, you can automatically perform a one-way analysis of variance F test to compare the means of two or more populations. For example, to perform a one-way analysis of variance F test to determine whether there is a difference in the mean flight times of Plane A, Plane B, and Plane C, do the following.

1. Use **SetUpEditor** to create and display three lists named TIMEA, TIMEB, and TIMEC in the stat list editor. Then enter the data into the appropriate list.

2. Verify using histograms or normal probability plots that the three samples are consistent with what you expect to obtain for samples of size 10 from normal populations.

3. Verify using **1-Var Stats** that it is reasonable to assume that the population standard deviations are equal. You should determine that the sample standard deviations for Planes A, B, and C are approximately 0.103, 0.110, and 0.147, respectively. Because the largest standard deviation is only about 1.4 times greater than the smallest standard deviation, the assumption of equal standard deviations is reasonable.

4. Perform an **ANOVA** test with the hypotheses: $H_0: \mu_1 = \mu_2 = \mu_3$ and $H_a: \mu_1, \mu_2, \mu_3$ are not all equal. To perform the test, enter the following keystrokes.

[STAT] [◀] [ALPHA] [F]	*Access **ANOVA** command.*
[2nd] [LIST] Choose TIMEA. [,]	*Paste list TIMEA.*
[2nd] [LIST] Choose TIMEB. [,]	*Paste list TIMEB.*
[2nd] [LIST] Choose TIMEC. [ )]	*Paste list TIMEC.*
[ENTER]	

After pressing [ENTER], the calculator will the display the F test statistic and the corresponding P-value. In this example, $F \approx 6.13$ and $P \approx 0.006$. These results provide strong evidence that you should reject H_0 and conclude that the planes do not all have the same mean flight lengths.

In addition to displaying the F test statistic and the corresponding P-value, the calculator will also display information concerning the variation among groups, the variation among observations in the same group, and the pooled standard deviation. The pooled standard deviation is an estimate of the common standard deviation, σ, of the observations in each group.

Classroom Exercises

1. Another way to calculate the P-value associated with the **ANOVA** F statistic is to execute the command

    Fcdf(6.13,1E99,2,27)

    from the home screen. To use **Fcdf,** press [2nd] [DISTR] [9] and enter the lower bound (in this case, the F test statistic), the upper bound (∞), and the degrees of freedom for the numerator (2) and the denominator (27). Execute this command to verify the activity results.

2. Use the results from the output of the **ANOVA** test to verify that the F test statistic is the ratio of the mean square for groups and the mean square for error.

3. Use the results from the output of the **ANOVA** test to verify that the pooled standard deviation s_p is the square root of the mean square for error.

Creating a Confidence Interval for μ_i

In this activity, you will learn how to use a TI-83 to create a confidence interval for μ_i where μ_i is an unknown mean from a population represented by one of the sample data sets. Consider the following problem.

> Using the flight time lengths for Plane A listed on page 153, compute a 95% confidence interval for the mean flight time of Plane A.

To create a confidence interval for μ_i, use the definition of confidence interval. By definition, the confidence interval for μ_i is

$$\bar{x}_i \pm t^* \left(\frac{s_p}{\sqrt{n_i}} \right)$$

where $\bar{x}_i$ is the sample mean of the ith sample, s_p is the pooled standard deviation, t^* is the critical value from the t distribution with degrees of freedom equal to the difference in the total number of sample values and the number of groups, and n_i is the sample size of the ith sample.

To create a 95% confidence interval for the mean flight length (time) of Plane A, do the following.

1. Determine $\bar{x}_i$, s_p, and n_i. From the results obtained in previous activities in this chapter, you know that $\bar{x}_i = 1.238$, $s_p \approx 0.121$, and $n_i = 10$.

2. Determine t^*. To determine t^* for 95% confidence and $30 - 3 = 27$ degrees of freedom, use the TI-83's equation solver to solve the equation

   $$0 = \text{tcdf}(-T,T,27) - .95.$$

   The resulting value is $T \approx 2.052$. (For more information on equation solver, see page 98.)

3. Determine the confidence interval by calculating the left and right endpoints. The left and right endpoints are

   $$\bar{x}_i \pm t^* \left(\frac{s_p}{\sqrt{n_i}} \right) \approx 1.238 \pm 2.052 \left(\frac{0.121}{\sqrt{10}} \right)$$
   $$\approx 1.238 \pm 0.0785$$

   The confidence interval is $(1.159, 1.317)$.

Classroom Exercises

1. Create a 95% confidence interval for the mean flight time of Plane B.

2. Create a 95% confidence interval for the mean flight time of Plane C.

3. In Chapter 18, page 110, the flight distances for two planes are listed. Use the distance data to perform an **ANOVA** F test to verify the results obtained in that activity. Specifically, verify that the values of the pooled standard deviation and the P-values are equal for both tests. Also, verify that the value of the F statistic is the square of the value of the t statistic.

A   Storing Data in Lists

The TI-83 has six standard lists in memory: L1, L2, L3, L4, L5, and L6. In addition to these standard lists, you can also define your own lists. Standard lists and user-defined lists can each contain up to 999 elements. The number of lists that can be stored in the TI-83 is limited only by the calculator's available memory.

You can arrange and display lists according to your own specifications using the stat list editor. The stat list editor enables you to store, edit, and view up to 20 lists at a time. To access the stat list editor, enter the following keystrokes.

|STAT| *Access stat edit menu.*
|ENTER| *Access stat list editor.*

You can navigate the stat list editor using the arrow keys.

In this appendix, you will learn how to use the TI-83 to

- enter data into standard lists.
- name a list.
- use programs to store lists.
- delete lists from memory.
- use programs to restore lists.

The following data represent the number of students who meet in three classrooms of Watts Hall during a school day. The class periods (1–8), the classrooms (Watts 301, Watts 306, and Watts 315), and the number of students in each classroom during each period are listed. You will use these data as you learn how to store data in lists.

CLASS PERIOD	WATTS 301	WATTS 306	WATTS 315
1	23	22	18
2	22	18	17
3	22	24	16
4	21	20	20
5	20	22	18
6	22	17	19
7	20	20	20
8	19	23	18

Table A.1

Entering Data Into Standard Lists

There are two primary ways to enter data into the standard lists: using the stat list editor and using the home screen.

Entering Data Using the Stat List Editor

Using the data from Table A.1, enter the class periods into L1 and the number of students in Watts 306 during each class period into L2 as described below.

Access the stat list editor and position the cursor on the first row of L1. Notice that L1(1) is displayed at the lower left corner of the viewing screen. To enter the class periods into L1, enter the following keystrokes.

1 [ENTER] 2 [ENTER] 3 [ENTER] 4 [ENTER]
5 [ENTER] 6 [ENTER] 7 [ENTER] 8 [ENTER]

> Lists L1 and L2 should be empty lists. If not, clear the lists by highlighting each list name and pressing [CLEAR] [ENTER].

You can enter the number of students in Watts 306 during each class period in a similar manner. Simply use the cursor arrows to position the cursor on the first row of L2, then enter the following keystrokes.

22 [ENTER] 18 [ENTER] 24 [ENTER] 20 [ENTER]
22 [ENTER] 17 [ENTER] 20 [ENTER] 23 [ENTER]

Entering Data Using the Home Screen

Using the data from Table A.1, enter the number of students in Watts 315 during each class period into L3 as described below.

Access the home screen. You can usually get to the home screen from any editor by pressing [2nd] [QUIT]. To enter the number of students who meet in Watts 315 during each class period into list L3, enter the following keystrokes.

[2nd] [{] 18 [,] 17 [,] 16 [,] 20 [,] 18 [,] 19 [,]
20 [,] 18 [2nd] [}] [STO▶] [2nd] [L3] [ENTER]

After pressing [ENTER], the list elements are stored in L3 and a portion of the list, followed by an ellipsis, is displayed on the home screen. The ellipsis indicates that the list extends beyond the viewing window. To see the remainder of the list, press [▶] repeatedly. If you go to the stat list editor, you can also view the elements of L3.

Appendix A

Naming a List

Naming a list is often convenient and helpful, especially if the data in the list are to be used again. List names can contain up to five characters. They must begin with a letter (A–Z) or the symbol θ. List names can also contain numbers, but not as the first character.

Storing Standard List Contents in a Named List

The names of the standard lists, L1, L2, etc., cannot be changed. Instead, you can use the home screen or the stat list editor to copy the contents of a standard list to a new user-defined list.

Using the Home Screen to Copy and Name a List

To copy the contents of L1 to a list named CLASS, enter the following keystrokes.

[2nd] [QUIT]	*Access home screen.*
[2nd] [L1] [STO▶]	*Access list to copy.*
[2nd] [ALPHA] [C] [L] [A] [S] [S]	*Name new list.*
[ENTER]	*Copy L1 to new list.*

[2nd] [ALPHA] turns alpha-lock on. When alpha-lock is on, only alphabet characters can be entered. You can turn alpha-lock off by pressing [ALPHA].

After pressing [ENTER], notice that the contents of L1 are copied to list CLASS, and the contents of CLASS are displayed on the home screen. To verify that CLASS is stored in memory, press [2nd] [LIST] to display the list names menu. Lists are displayed in the list names menu in alphabetical order. If this is the first time you have named a list, CLASS will be stored as the first list. Note that lists L1 through L6 are not in the list names menu. These lists are accessed directly from the calculator keyboard.

To insert and display list CLASS in the stat list editor, enter the following keystrokes.

[STAT] [ENTER]	*Access stat list editor.*
Highlight L1.	*Highlight L1.*
[2nd] [INS]	*Insert new list.*
[C] [L] [A] [S] [S]	*Type name after Name=.*
[ENTER]	*Insert and display list.*

After pressing [ENTER], notice that the name CLASS and its contents are inserted and displayed at the left of L1.

Storing Data in Lists   A3

A more efficient method for inserting a list into the stat list editor is to paste the list name instead of typing it. For example, copy the contents of L2 into a list named W306. Then insert and display the contents of W306 in the stat list editor using the following keystrokes.

[STAT] [ENTER]   *Access stat list editor.*
Highlight L1.   *Highlight desired list.*
[2nd] [INS]   *Insert new list.*
[2nd] [LIST]   *Access list names menu.*
Choose W306.   *Identify and paste desired list.*
[ENTER]   *Insert and display list.*

> To choose a list, cursor to the list name and press [ENTER].

Using the Stat List Editor to Copy and Name a List

To copy the contents of list L3 to a new list named W315, enter the following keystrokes.

[STAT] [ENTER]   *Access stat list editor.*
Highlight L1.   *Highlight desired list.*
[2nd] [INS]   *Insert new list.*
[W] [ALPHA] 3 1 5 [ENTER]   *Name list.*
[2nd] [L3] [ENTER]   *Copy L3 to W315.*

> Alpha-lock turns on automatically when using [2nd] [INS] to insert a new list. Press [ALPHA] to turn alpha-lock off.

Notice that list W315 is automatically displayed.

Naming and Creating a New List

Using the data from Table A.1, you can name, create, and display a new list called W301 by entering the following keystrokes.

[STAT] [ENTER]   *Access stat list editor.*
Highlight W306.   *Highlight desired list.*
[2nd] [INS]   *Insert new list.*
[W] [ALPHA] 3 0 1 [ENTER]   *Name list.*
[-] 23 [ENTER] 22 [ENTER]   *Enter data.*
22 [ENTER] 21 [ENTER]
20 [ENTER] 22 [ENTER]
20 [ENTER] 19 [ENTER]

Note that the new list is inserted between lists CLASS and W306. To verify that the named lists are stored in memory, press [2nd] [LIST].

A4   Appendix A

Additional List Editing Tips

- Editing a list is often necessary when correcting or updating data. To change a data entry in a list, simply highlight the entry to be changed, type the new entry, and press [ENTER].

- To remove a list from the stat list editor, position the cursor to highlight the name of the desired list and press [DEL]. The list is no longer displayed; however, it is still stored in memory.

- To delete the entire contents of a list, position the cursor to highlight the name of the desired list and press [CLEAR] [ENTER]. The list name will remain in the stat list editor, but its entire contents are erased from memory. Note the difference between using the [DEL] and [CLEAR] keys while in the stat list editor. Be very careful when using the [CLEAR] key.

Using Programs to Store Lists

Storing too many lists in memory will exhaust the TI-83's memory capacity. Because the TI-83 uses less memory to store a program than it does to store a list, it is more efficient to store a list (or lists) in a program.

The first step in writing a program is to choose an appropriate program name. Program names can contain up to eight characters. When choosing a program name, choose a name that identifies the data stored in the program. For example, storing the data contained in lists CLASS and W301 in a program called WATTS will help you identify the data if you need to use it at a later time.

To write program WATTS, enter the following keystrokes.

> The designation "LCLASS" identifies the string CLASS as a list.

[PRGM] [▶] [▶] [ENTER]	*Create new program.*
[W] [A] [T] [T] [S]	*Define program name.*
[ENTER]	*Access program editor.*
[2nd] [RCL] [2nd] [LIST]	*Recall CLASS into program WATTS.*
Choose list CLASS.	
[ENTER]	
[STO▸] [2nd] [LIST]	*Store data to list CLASS.*
Choose list CLASS.	
[ENTER]	
[2nd] [RCL] [2nd] [LIST]	*Recall W301 into program WATTS.*
Choose list W301.	
[ENTER]	
[STO▸] [2nd] [LIST]	*Store data to list W301.*
Choose list W301.	
[ENTER]	

> To exit the program editor and go back to the home screen, press [2nd] [QUIT].

Storing Data in Lists   A5

Deleting Lists from Memory

The primary reason for writing a program to store lists is to save memory. Therefore, you should delete appropriate lists as soon as you write such a program. The following keystrokes show how to delete lists CLASS and W301 from memory.

[2nd] [MEM]	*Access memory menu.*
[2]	*Access delete menu.*
[4]	*Access list menu.*
Choose list CLASS.	*Delete list.*
Choose list W301.	*Delete list.*
[2nd] [QUIT]	*Exit delete list menu.*

Press [2nd] [LIST] to verify that lists CLASS and W301 are no longer stored in the calculator's memory.

Using Programs to Restore Lists

To restore lists that have been stored in a program, simply execute the desired program. To execute a program, press [PRGM] and choose the appropriate program name from the EXEC menu. You can restore lists CLASS and W301 by entering the following keystrokes.

[PRGM]	*Access program EXEC menu.*
Choose program WATTS.	*Paste program name to home screen.*
[ENTER]	*Execute program.*

To verify that lists CLASS and W301 are restored to memory, press [2nd] [LIST] to view the list names menu. Notice, however, that executing the program does not restore the lists to the stat list editor.

B Using SetUpEditor

In this appendix, you will learn how to use the TI-83's **SetUpEditor** command. Using **SetUpEditor,** you can

- restore the stat list editor to its default setting.
- define the stat list editor display.
- create and display new lists.

The following sections use the data from Table A.1 of Appendix A. It is assumed that the data are stored in your calculator.

Restoring the Stat List Editor

By default, the stat list editor displays the six standard lists L1, through L6. If you have altered the stat list editor so that it displays other lists, you can easily restore the stat list editor to its default setting by entering

[STAT] [5] [ENTER].

After entering these keystrokes, you should notice the **SetUpEditor** command and a Done message displayed on the home screen.

To verify that **SetUpEditor** works, do the following.

1. Access the stat list editor and insert a named list between L3 and L4.

2. Delete each of the standard lists, L1 through L6, by highlighting each list name and pressing [DEL]. After deleting the standard lists, the stat list editor should display the named list and two empty columns.

3. Restore the stat list editor to its default setting as described above.

4. Access the stat list editor to view the results. Note that any data stored in the standard lists before you deleted the lists still exist.

Defining the Stat List Editor Display

Using **SetUpEditor,** you can set up the stat list editor to display lists stored in memory. For example, to display list W306 in the stat list editor, enter the following keystrokes.

[STAT] [5] *Select* **SetUpEditor** *command.*
[2nd] [LIST]
Choose W306. *Paste desired list.*
[ENTER]

After pressing [ENTER], notice the Done message displayed on the home screen. Verify that the stat list editor displays list W306 and two empty columns. Note that although the standard lists are not displayed, each list and its contents are stored in memory.

Creating and Displaying New Lists

In addition to using **SetUpEditor** to display lists stored in memory, you can also use **SetUpEditor** to create and display new lists. For example, to set up the stat list editor to display lists CLASS, W301, W306, W315, and a new list named TOTAL, enter the following keystrokes.

[STAT] [5] *Select* **SetUpEditor** *command.*
[2nd] [LIST] Choose CLASS. *Paste desired list.*
[,] [2nd] [LIST] Choose W301. *Paste desired list.*
[,] [2nd] [LIST] Choose W306. *Paste desired list.*
[,] [2nd] [LIST] Choose W315. *Paste desired list.*
[,] [2nd] [ALPHA] [T] [O] [T] [A] [L] *Name new list.*
[ENTER]

After pressing [ENTER], notice the Done message displayed on the home screen. Verify that the stat list editor displays lists CLASS, W301, W306, W315, and TOTAL. Note that as you navigate through the stat list editor, a number is displayed in the upper right corner of the viewing window. This number denotes which one of the 20 available columns is occupied by the currently highlighted list. List TOTAL should occupy the fifth column. You should also verify that list TOTAL is stored in the list names menu.

C Defining Lists Using Formulas

Using formulas, you can define a list in terms of other lists. In this appendix, you will learn how to use the TI-83 to

- define a list using an unattached formula.
- define a list using an attached formula.
- edit an attached formula.
- detach a formula from a list.

The following sections use results obtained by completing the sections in Appendix B. It is assumed that the correct data are stored in your calculator.

Defining Lists Using Unattached Formulas

Defining a list using an unattached formula allows you to create a list using lists stored in memory and then edit one or more of the stored lists without changing the new list. Lists defined using an unattached formula are editable.

For example, using the data in Table A.1 of Appendix A, you can use an unattached formula to define a list named TOTAL that represents the total number of students assigned to classrooms W301, W306, and W315 each period throughout the school day. There are two ways to define a list using unattached formulas: using the home screen and using the stat list editor.

Using the Home Screen

To define list TOTAL using the home screen, enter the following keystrokes.

[2nd] [QUIT]	*Access home screen.*
[2nd] [LIST] Choose W301.	*Paste desired list.*
[+] [2nd] [LIST] Choose W306.	*Paste desired list.*
[+] [2nd] [LIST] Choose W315.	*Paste desired list.*
[STO▸] [2nd] [LIST] Choose TOTAL.	*Store to desired list.*
[ENTER]	

After entering the keystrokes, notice that the home screen displays the contents of list TOTAL. The first element in the list represents the total number of students in the three classrooms during first period. That is,

$$TOTAL(1) = W301(1) + W306(1) + W315(1).$$

Using the Stat List Editor

To define list TOTAL using the stat list editor, first clear any entries in the list. Then, enter the following keystrokes.

Highlight TOTAL.	*Highlight list to be defined.*
[2nd] [LIST] Choose W301.	*Paste desired list.*
[+] [2nd] [LIST] Choose W306.	*Paste desired list.*
[+] [2nd] [LIST] Choose W315.	*Paste desired list.*
[ENTER]	

After entering the keystrokes, notice that the cursor is positioned on the first element of list TOTAL and the entry line at the bottom of the screen displays TOTAL(1) = 63. The entries of list TOTAL are identical to those obtained when defining the list using the home screen.

> After defining list TOTAL, verify that it is defined using an unattached list. To do so, change an entry in list W301, W305, or W315. Does list TOTAL change?

Defining Lists Using Attached Formulas

Defining a list using an attached formula also allows you to create a list using lists stored in memory. The difference between defining a new list using an attached formula and an unattached formula is that when changes are made to any list used in the attached formula, the new list is automatically updated. Lists defined using attached formulas are not editable.

You can define a list using an attached formula from the home screen or the stat list editor. Either way, the procedure is very similar to the procedure for defining a list using an unattached formula. The only difference is that when defining a list using an attached formula, the formula is enclosed in quotes. To obtain a quotation mark, press [ALPHA] ["].

For example, to define list TOTAL using an attached formula, do one of the following.

- From the home screen, enter the formula

    "LW301 + LW306 + LW315" → LTOTAL.

- From the stat list editor, highlight list name TOTAL, then enter the formula

    "LW301 + LW306 + LW315".

> Verify that list TOTAL is defined using an attached formula by changing an entry in list W301, W305, or W315. What happens to the corresponding entry in list TOTAL?

Regardless of the method used, the results are the same. Notice the formula-lock symbol next to list TOTAL in the stat list editor. This symbol denotes a list defined by an attached formula.

> Verify that list TOTAL is not editable by trying to change one of its entries.

Editing Attached Formulas

To make changes to a list defined by an attached formula, you can edit the formula that defines the list. To edit an attached formula, first access the stat list editor and highlight the desired list name to display the formula in the entry line. Then, press [ENTER] to position the cursor on the entry line and edit the formula as necessary.

For example, to delete the data in list W306 in list TOTAL, highlight list TOTAL in the stat list editor. Press [ENTER]. Position the cursor over the "L" preceding W306 and press [DEL]. Repeat for the characters W, 3, 0, 6, and +. Then press [ENTER] again. Note how the entries in list TOTAL change.

Detaching Formulas From Lists

There are four ways to detach a formula from a list. Each method produces different results. You can detach a formula from a list

- using the home screen.
- using the stat list editor.
- by editing the list indirectly.
- by clearing the list.

Using the Home Screen

To detach a formula from a list using the home screen, enter

[ALPHA] ["] [ALPHA] ["] [STO▶] *listname* [ENTER].

No message is displayed on the home screen, but when you return to the stat list editor you will see that the formula-lock symbol is no longer displayed. Detaching a formula from a list using the home screen will not affect the list contents.

Using the Stat List Editor

To detach a formula from a list using the stat list editor, first highlight the list name. The formula used to define the list should be displayed in the entry line. Press [ENTER] to position the cursor on the entry line. Then press [CLEAR] [ENTER] to detach the formula. After pressing [ENTER], notice that the formula-lock symbol disappears and that the list entries are unchanged. Notice also that if you highlight the list name again, the entry line displays the elements at the beginning of the list.

Editing the List

Lists defined using an attached formula are not directly editable. You can edit such a list indirectly, however, but doing so results in detaching the formula. To edit a list element and simultaneously detach the formula, first position the cursor over the element to be changed. Notice that the element appears in the entry line without the cursor. Press [ENTER] and notice that the cursor appears. Change the entry as desired and press [ENTER] again. The formula is no longer attached to the list and the formula-lock symbol disappears.

Clearing the List

A more destructive way to detach a formula from a list involves using the **ClrList** command. To detach a formula from a list using **ClrList,** first select the command by pressing [STAT] [4]. Then paste the desired list name after the **ClrList** command and press [ENTER]. A done message is displayed on the home screen. When you return to the stat list editor, the list name remains but all its contents are deleted.

> Use one of the methods discussed in this section to detach the formula from list TOTAL.

Note: Displaying a "locked" list in the stat list editor slows the data entry and editing when working with other lists. This is the case even if the list with which you are working is not part of the formula. To speed the editing and data entry process, you should either detach the formula or rearrange the stat list editor display so that no "locked" lists are displayed.